The Green Skyscraper

The Green

Sky

scraper

The Green
Skyscraper

The Basis for
Designing Sustainable
Intensive Buildings

Ken Yeang

Prestel
Munich · London · New York

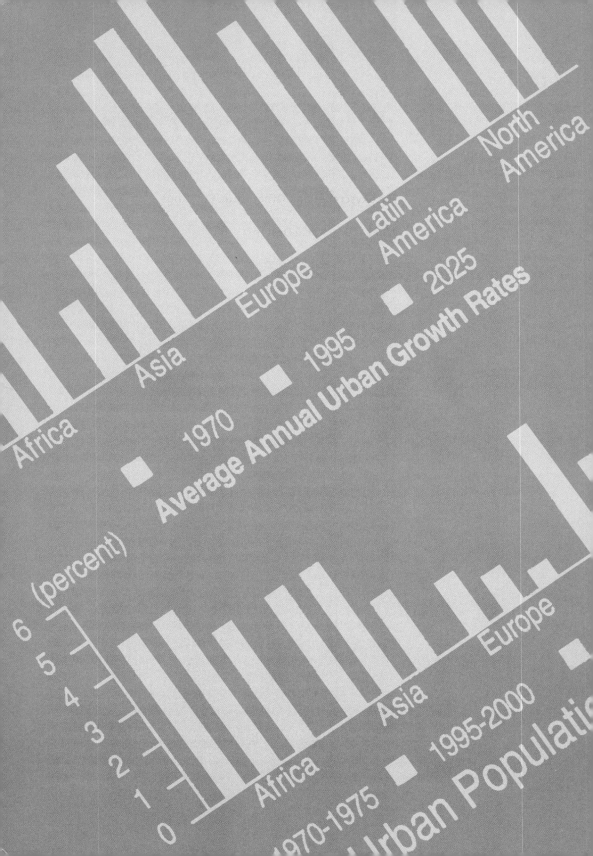

Average Annual Urban Growth Rates

Africa
Asia
Europe
Latin America
North America

■ 1970
■ 1995
■ 2025

(percent)

6
5
4
3
2
1
0

Africa
Asia
Europe

■ 1970-1975
■ 1995-2000

Urban Population

Contents

This book is intended as a primer for the ecological design of large buildings, particularly the skyscraper or the tall building type.

There are three tenets that motivate the publication of this book. The first is that ecological design or sustainable design is, at its most basic level, an approach that seeks to ensure that future generations enjoy continued access to natural resources. Second, it is here contended that building designers (the architects, engineers and other specialists involved with the production of buildings) can make a significant difference and can contribute to enabling the achievement of a sustainable future. Lastly, I believe that the skyscraper and other large buildings deserve greater attention in terms of 'green' design.

It has already been well argued elsewhere that designers should make our manmade environment ecologically sustainable or 'green' (e.g. WCED, 1987; Sitarz, D., 1994) (see Fig. 1). Many are already well aware of the extensive degradation inflicted on the natural environment; for instance, the earth's biodiversity is calculated to be degraded at the rate of 50,000 species per annum (Brown, 1991) as a result of humanity's callously destructive activities. With regard to the earth's inorganic resources, it is also clear that the current relatively low-cost provision of energy to the built environment from non-renewable sources and the profligate use of irreplaceable materials can certainly not continue (see Fig. 2). It is projected that, at most, we might expect to have a sufficiency of biospheric non-renewable energy resources for perhaps another 50 years (Von Weizsacker, Lovins, A.B., and Lovins, L.H., 1997). It is therefore evident that designing with 'green' or ecologically responsive design objectives in mind is vital. Indeed, these must certainly now be the prime objectives for the design community today.

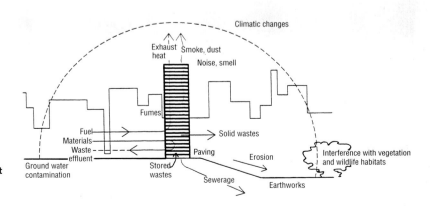

Fig. 1 Built environment impacts on its surrounds (Source: Yeang, 1995)

"Green" or ecological design here means building with minimal environmental impacts, and, where possible, building to achieve the opposite effect; this means creating buildings with positive, reparative and productive consequences for the natural environment, while at the same time integrating the built structure with all aspects of the ecological systems (ecosystems) of the biosphere over its entire life cycle.

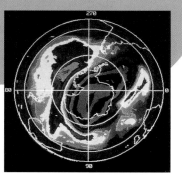

This is the primary thrust of this book, for in the event that we fail to achieve these objectives, our widespread activities as a species on the earth will in time overload the resilience (or the carrying capacities) of the other species and the natural systems of our planet, inevitably leading to total devastation of the natural environment, and eventually of our own built environment – our world – as a consequence. It is worth keeping in mind that the human species is growing by more than 250,000 people per day.

It is evident that it is the early stages in the production of a building, especially at the design stage, that offer the greatest opportunities for addressing the anticipated problems of environmental impairment that may arise later in the building's life cycle. More than 45 percent of energy used in a country goes to buildings, and up to 26 percent of landfill waste comes from building construction (see Fig. 3). In this regard, the architect and designer of buildings can most certainly contribute significantly to a sustainable future. Important early design decisions start when the designer selects the raw and assembled materials to be used in the building, the type of operational systems for the building, and the routes to be taken to the final 'sink' for all the materials, parts and waste energy from the built system. All these decisions will have a greater or

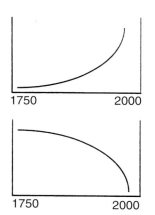

Fig. 2   Rates of change on the planet (Source: Loening, V., 1992, in Reid, 1995)

On the top: *Increases in human population, consumption of fossil fuel, consumption of other resources, industrial production, agriculture, desertification, salinisation, pollution, military expenditure.*

On the bottom: *Decreases in forests, fish stocks, farmland, soils, habitats, species, biodiversity, environmental services, human diversity.*

Fig. 3   Building inputs
and outputs (Source:
Yeang, 1998)

lesser environmental impact depending on whether they are made as preventive decisions or as environmentally callous decisions. Of course, we have to recognise the fact that in ecological design there will certainly be no such thing as a perfect solution, nor will there be instant technological fixes that will solve the myriad of environmental issues associated with producing a building, particularly at the scale of a skyscraper. Neither are there hard and fast rules for all our design endeavours.

Building inputs
• 40 %   of **raw materials** (by weight) are used in building construction globally each year
• 36–45 %   of a nation's **energy** input is used in buildings

Building outputs
• 20–26 %   of landfill trash is construction trash
• 100 %   of **energy** used in buildings is lost to the environment

In green design, the ecological responsiveness of a particular design solution depends almost entirely upon the extent of the designer's ability and ingenuity. This is simply because green design is a process in which the environmental attributes of a building are treated as objectives rather than as constraints (Goldbeck, The Office of Technology Assessment, U.S. Congress, 1995). Therefore, in most instances, a design's success in fulfilling its ecological objectives depends much upon the creativeness of the designer and the inventiveness of the solutions he or she generates to address the limitations and opportunities presented by the site and in the design problem itself. Simply stated, the greater the adherence to the principles of applied ecology, and in particular, those that are outlined in chapters 2 to 6 of this book, the greater will be the effectiveness of the ecological solution.

For many designers today, environmentally responsive objectives are already firmly considered to be the prerequisites of all design endeavours. More than just accepting ecological design objectives as the prerequisites for their work, however, architects must exert a far more decisive influence on the conception and layout of buildings, particularly with regard to their content, siting, materials, operational systems and other features.

What is also distressing is that many of the ecological designers today still prefer to deal primarily with small-scale buildings (i.e., with low-rise and medium-rise buildings) and often only in greenfield and rural sites (e.g., Crosbie, M.J., 1994). All those large-scale and intensive building types (such as skyscrapers) located in urban areas and in dense inner cities are regarded as anathema. Undoubtedly, intensive urban buildings are indeed energy guzzlers, consum-

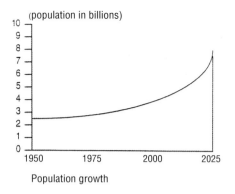

(population in billions)

Population growth

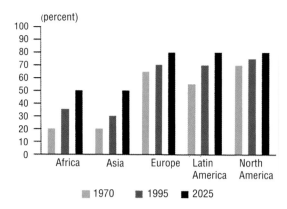

1970   1995   2025

ing huge amounts of materials in their
construction and making massive volumes
of waste discharge into the environment.
This disparaging view of these building
types is of course not disputed here, but
surely these are precisely the reasons why
such buildings demand our attention. For
if all ecological designers refuse to con-
front the design of intensive large urban
building, then who will? Furthermore, it
may be argued that it is these very high-
density intensive buildings that should command by far the greater
part of our expertise and effort with regard to creating ecologically
healthy and responsible designs than the smaller buildings which
present fewer problems.

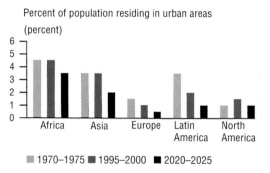

Percent of population residing in urban areas

1970–1975   1995–2000   2020–2025

Fig. 4   Average annual urban growth rates (Source: UN, New York, 1995)

The myth of the idyllic, self-sufficient, decentralised and aut-
archic community located in the countryside as the panacea for
human settlements remains, unfortunately, a myth. Whether we
like it or not, the cities around the globe will most certainly continue to expand and to develop with increasing intensity. Intensive
urban structures such as skyscrapers will certainly continue to be
built, unless of course an alternative viable built form presents
itself as a substitute, or if there are large-scale changes to the existing urban land economy or arising from the digital revolution.

Furthermore, despite the changes and the benefits of the digital
revolution, the rural-to-urban migration of the population in the
developing world will continue to increase. New cities and new
urban settlements will most certainly proliferate, many almost
overnight.

As a result of these trends, the building of intensive urban developments of all types will most surely continue. Skyscrapers, though

hateful they may be to some, will continue to be built as long as land prices continue to go up and entrepreneurs conclude that the only way to recoup the high cost of urban land is to increase plot ratios – i.e. to build upwards and more intensively on a given piece of land (see Fig. 4). This is unfortunately the reality of the day, an inescapable fact of the economics of urban land economy.

On the other hand, some might be tempted to claim on the basis of these same facts that skyscrapers are indeed the preferred green building form over the low-rise precisely because they have a smaller footprint, and the residual portion of the land (because of the lower land coverage) can be returned to nature (see chapter 1). Therefore, it is held here that the skyscraper and similar intensive building types need immediate and urgent attention from the world's ecological designers to make sure that these are built to be ecologically responsive, or at least as environmentally responsive as possible for our sustainable future.

We have to of course accept that many of the problems and the technical innovations for a comprehensive holistic ecological design for intensive building types remain unresolved or have yet to be invented. But this should not lead us to assume that a technological 'fix' is the preferred solution of design problems or that it is possible for all environmental issues to be resolved overnight.

What we must do is to make an intelligent start, based on current knowledge, on the development of the techniques and ideas needed to solve green design problems; it is hoped that this book will provide the foundation upon which such endeavours can be based.

To avoid confusion between what is bioclimatic design and what is ecological design, we should clarify the differences. Generally, bioclimatic design is the passive low-energy design approach that makes use of the ambient energies of the climate of the locality to create conditions of comfort for the users of the building. Initially, we employed bioclimatic design principles in earlier work on the design of the tall large building type to produce an alternative form to that of the conventional skyscraper. This kind of structure performs differently from the conventional skyscrapers in the first instance as a passive low-energy configuration. As an emergent bioclimatic built form, it provides a viable alternative to the existing skyscraper and constitutes a new building genre; however, it must be made clear that bioclimatic design is not ecological design in its entirety, but only an intermediate stage in that direction. Ecological design is a much more complex endeavour and should be differentiated clearly from the design approaches of other architects (see Fig. 5).

The ideas and theories in this work should be distinguished from other expressions of 'ecological design'. The emphasis here is on the interdependencies and interconnectedness in the biosphere and its ecosystems. It is contended here that the crucial property of ecological design is the connectedness between all activities, whether manmade or natural; this connectedness means that no part of the biosphere is unaffected by human activity and that all actions affect each other. This property is made explicit and inescapable in the theoretical 'Interactions Matrix' in chapter 3 and the 'law of ecological design'. Simply stated, all built systems must have a reciprocal relationship with their local environments and with the rest of the biosphere (Behling & Behling, 1996, p. 235).

This property of interconnectedness is absent in the ecological design theory and practice of a large number of architects. For instance, some architects define ecological design (or ESD – 'Envir-

Fig. 5 Design mode (others, bioclimatic & ecological) (Source: Yeang, 1995)

	Design mode		
	Others	Bioclimatic	Ecological
Built form configuration	Other influences	Climate influenced	Environment influenced
Building orientation	Relatively unimportant	Crucial	Crucial
Facade and windows	Other influences	Climate responsive	Environment responsive
Energy source	Generated	Generated / ambient	Generated / ambient / local
Energy loss	Relatively unimportant	Crucial	Crucial / reused
Environmental control	Electro-mech	Electro-mech / manual	Electro-mech / manual
	Artificial	Artificial / natural	Artificial / natural
Comfort level	Consistent	Variable / consistent	Variable / consistent
Low-energy response	Electro-mech	Passive / electro-mech	Passive / electro-mech
Energy consumption	Generally high energy	Low energy	Low energy
Materials source	Relatively unimportant	Relatively unimportant	Low environmental impact
Materials output	Relatively unimportant	Relatively unimportant	Reuse / recycle / reintegrate
Site ecology	Relatively unimportant	Important	Crucial

onmentally Sustainable Design') as ' . . . minimal energy use, minimal water use, minimal waste, maximising human health, promoting biodiversity'. This definition is in part correct: all these factors
do contribute to ecological design. But if the interdependencies and
connectedness between these factors and the natural systems in
the biosphere are not considered, then this approach is incomplete
and hence not ecologically correct, and it could even be ecologically
hazardous.

It should also be added that while examples of skyscraper
designs and building operational systems are shown or discussed
here, we must at the same time recognise that ecological design is
still in its early days, and that none of the technical examples
shown (e.g. in chapter 6) can claim to be completely 'green', nor are
they panaceas. In fact, there will inevitably be aspects of many of
these systems that could be criticised. Nevertheless, the examples
shown may have some aspects that demonstrate or illustrate a
principle, device, innovation or feature that is relevant to the
advancement of green design. However, it is clear that the tenta-
tiveness of existing technical solutions should not in any way invalidate the theory presented here (in chapter 3).

More important, it is hoped that this primer will redirect designers away from the worst-case, energy-profligate building type,
which has been referred to as the "iced-tea" building (Fordham,
1997), and turn them towards a more ecologically responsible building form. The proposition of the "iced-tea" building is a ridiculous
one. In this analogy, energy is used to boil water to make the tea.
More energy is then used to chill the tea before we drink it. Similarly, the "iced-tea" building is one that is energy inefficient (for example, the all-glass tower), and spends a good part of the day in the
summer absorbing and collecting the heat of the sun, making it
extremely hot, and then consumes more non-renewable energy (i.e.
through air-conditioning) to cool itself down. Transforming such
buildings from "solar-ovens" into "walk-in electric refrigerators"
uses about 16 percent of the electricity in the United States through
air-conditioning systems (Von Weizsacker et al., 1997, p. 58).

In this field of endeavour where experimentation remains the
order of the day, it is too easy for detractors to berate the discrepancies between the theoretical writings and actual buildings. Let
me be the first to admit this, and to explain. The discrepancy between theory and practice arises partly because buildings are often
designed several years before they are completed on-site (one of the
hazards of architectural practice), whereas the ideas, theoretical
development and writings may have advanced considerably in the
interim. As a result, divergences between built works and the cur-

rent status of theoretical and technical thinking are inevitable and will always be prevalent. We of course must always try to make sure that our subsequent works make up for previous discrepancies, but they will no doubt exhibit further discrepancies upon the publication of newer writings.

Having acknowledged here that ecological design is still in its infancy, we must recognise that many of the technical solutions discussed may not be totally rigorous nor entirely acceptable. For example, the embodied ecological impact of the production and recovery after use may not have been adequately evaluated. Furthermore, the usual predicaments of limited time and limited budget, together with a host of other regulatory and code restrictions, will continue to inhibit the adequate resolution of all pressing or relevant ecological issues within a specific design project. This is unfortunately one of the vicissitudes of architectural practice in a commercial environment. Clearly, much still remains to be invented or resolved by creative design means. Current ecological design strategies should be more appropriately regarded as a transition towards the ecological ideal.

The inadequacies of existing technical solutions and perceived discrepancies between theory and practice should certainly not invalidate the theoretical structure, provided that the latter is based on fundamentally correct premises. Neither should the perceived discrepancies invalidate the design of completed buildings, provided that these are regarded simply as 'rapid prototypes' (or 'imperfect working models', the physical manifestation of ideas). Buildings are often produced 'on the run' in a rapid design-and-build commercial environment, and many buildings are simply progressive experimental prototypes – albeit achieving less than one might expect otherwise because as mentioned earlier the real-world conditions are dissimilar to ideal laboratory conditions.

We should also be clear that the purpose of this treatise here is not to provide a step-by-step guide for the design of the green skyscraper or larger buildings. The design and coordination of all the technical, aesthetic, spatial, structural and environmental engineering factors inherent in large building design are really too extensive to discuss here in their entirety. A design primer with too much prescriptive detail looses its usefulness and flexibility. Worst, it may inhibit creativity. It is assumed that the reader has already the fundamental understanding of how to design larger buildings such as the skyscraper (see Yeang, 1996) and that this work serves to enable him or her to review, modify and revise his or her own approach to skyscraper design in the light of the ecological considerations discussed here.

It is hoped here that by setting out the relationship between key design decisions and ecological issues, this work will help the designer of the skyscraper or other large building types to ask the right questions in the design process and make the appropriate informed decisions, which will then demand appropriate technical solutions.

If the appropriate technical solution is not in existence, it is hoped that this primer will lead to the invention of new solutions and the creation of new responses that will further advance this field of endeavour.

Ecological design, as will be made evident in subsequent chapters here, includes not just architectural and engineering design but also other seemingly disparate disciplines such as landscape ecological land-use planning, embodied energy studies, recycling practices, pollution control, etc. The value of this treatise is the bringing together and integration of these aspects of environmental protection and control (previously regarded as separate disciplines) brought into a single approach to ecological design (see chapter 3, partitioned matrix).

Notes on Terminology

● The term 'skyscraper' is used throughout this book as a convenient abbreviation for the large high-rise intensive building type, generally regarded as being over 10 storeys and which can be of commercial, residential, hotel or mixed use.
● For expediency, the term 'man' is used here to mean 'human' and is not gender-specific.
● The term 'designer' refers to all those involved with the design of the built environment, whether they are architects, designers, engineers, planners, builders or other professionals. Similarly, in cases where the word 'architect' is used, this is not meant to be restrictive; the ideas discussed here are offered for the consideration and use of all those involved or interested in the design of green skyscrapers and other intensive building types.

Ken Yeang   1999

Chapter One

The Skyscraper and
Other Large Buildings

Intensive and High-Density Urban Contexts

In developing a green design theory for the tall building type, a logical first question is, why skyscrapers and large buildings? If designing in an ecologically responsible way is already such a huge and complex endeavour, surely it is foolhardy, one might think, to complicate the task further by seeking to design these green principles into large and complex buildings located in intensive, high-density urban contexts such as are found in our cities.

My contention is simply that the issue of the ecological design of large buildings (whether we like these buildings or not) is just as vital as the ecological design of the smaller building types – in fact more crucial, because of their scale and volume of consumption of energy and materials. The massive scale of such buildings often means that issues are not addressed or are negated because they appear daunting and unmanageable (see Fig. 1), especially to the inexperienced or uninformed designer. The skyscraper or high-rise built form is a specialised building type whose design requires an acquired expertise. For example, it requires special engineering systems because of its height, and this demands particular design attention.

The most obvious reason why we should develop a green design has already been alluded to in the preface: the skyscraper (whether for office, apartment, hotel or other use) is probably the most ubiquitous building type in all major cities today. Whether we like it or not, the skyscraper as a building type is here to stay. It will not just go away because many do not like it or find it hateful. Unless there are radical changes in current business practices (e.g. in the extensive use of digital telecommunications systems) or in urban transportation, or if there is a reversal of the urban-to-rural migration trends, the skyscraper will continue to be built in large numbers well into the first quarter of the millennium. This claim is supported by projections that by the year 2000, 47 percent of the world's population will be urbanised, and 24 cities in the world will exceed 10 million inhabitants. In economic models of clustering, transportation costs and labour mobility would have a central role: if transportation is cheap and labour is mobile, agglomerations will tend to outweigh dispersal (and the converse). With such intensification of urban living, the impact upon the atmosphere and ecosystems will become critical issues facing urban planners and designers well into the first part of the next millennium (Steele, 1997).

Cities and their large buildings demand greater attention with regard to their ecological design because these are the places where the problems of resource-consuming and environment-polluting economic relations and ways of life threaten global natural resources and ecosystems most clearly and most insistently. The chances of achieving a global policy for a sustainable future will likely be decided in the cities. If projections of rural-to-urban migration prove correct, then the key building type that architects will need to pay serious attention to may well be high-density urban buildings such as skyscrapers. Spatially, it is predicted that by year 2006, there will be half a billion people inhabiting 1 percent of the earth's land surface (Fazal, 1995). The obvious consequence of such compression of large numbers into small spaces is to build upwards to accommodate the swelling urban population.

There are, of course, other intensive urban building types besides the skyscraper, such as shopping malls, museums, supermarkets, etcetera. However, most of these tend to be medium-rise buildings, unless they are in a composite form such as podium-and-tower. The design issues for medium-rise buildings are already well covered elsewhere, but not those of the skyscraper.

Can the Skyscraper be 'Green'?

The proposition that the skyscraper and other high-density intensive building types can be designed to be ecologically responsible may be regarded by some with great suspicion. For those to whom skyscrapers are energy-hungry parasites that feed upon the city's surrounding ecosystems, the landscape and global resources, a 'green' skyscraper may be a contradiction in terms. Or one might compare the skyscraper to considerably smaller, experimental low-energy ecological buildings, and find the latter to be the self-evident ecological building type of the future. The reasoning is that by virtue of their enormous size, skyscrapers consume huge amounts of energy and materials and make similarly extensive discharges into the natural environment (a charge not denied here), and are thus inherently un-green.

Such views often fail to look at the entire life cycle of a building (a point that will be underscored frequently in this book) and the

The green urban centre

larger web of interrelated human and environmental systems. In fact, it is in the case of the skyscraper in particular that the eventual recovery of the building's materials and components at the end of their life is most significant (more so than, say, for smaller building types, where recovery may be less financially justifiable) simply because of its scale. Thus, the skyscraper offers the greatest possibilities for recycling of precious resources. As mentioned earlier, there are also a number of positive justifications for the skyscraper's existence. The main ones are planning considerations that affirm the skyscraper as an effective green alternative to the well-known low-rise, decentralised suburban structure. Again, failure to look at the larger picture and the totality of system interactions distorts comparisons and makes the case for the skyscraper look worse than it is, for the decentralised form of built environment (as opposed to the intensified urban solution) requires higher consumption of non-renewable energy resources, particularly for transportation. Obviously, a decentralised planning layout of structures means further travel distances between buildings. Recent studies have shown that the greater the intensification of urban population, the lower is the energy consumption per inhabitant for travel in automobiles. Indeed, there appears to be a geometrical relationship in the reduction of energy consumption through transportation to the increase in building density (see Fig. 6).

The transportation issue provides a crucial justification for dense building development, for the existence of cities and for intensive urban buildings such as the skyscraper. Furthermore, this lower-energy justification applies not just to horizontal transportation systems but to vertical transportation systems as well. For example, elevators in skyscraper are estimated to be 40 times more energy efficient and 10 times more materials efficient than an average 1995 automobile (Von Weiszacker et al., 1997, p. 94). But it is only when we consider how the building is used, and not just what it is, that these environmental and energy costs become clear. Automobiles consume land in highways, and in feeder roads and parking areas that are entirely dedicated to them. They also consume the

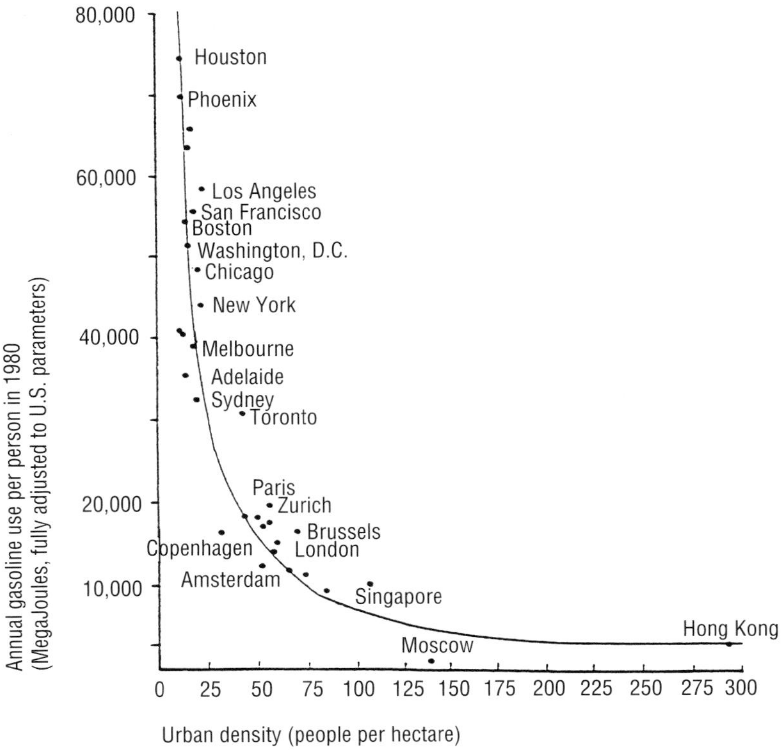

Fig. 6   Gasoline consumption and urban densities (1980) (Source: Robert Paehlke, 1989)

Annual gasoline use per person in 1980
(Megajoules, fully adjusted to U.S. parameters)

80,000

Houston

Phoenix

60,000

Los Angeles
San Francisco
Boston
Washington, D.C.
Chicago

New York

40,000

Melbourne
Adelaide
Sydney
Toronto

Paris
20,000   Zurich
Brussels
Copenhagen   London
Amsterdam
10,000
Singapore
Hong Kong
Moscow

0   25   50   75   100 125 150 175 200 225 250 275 300

Urban density (people per hectare)

majority of the street space that was originally given to pedestrians. Of urbanised land, 25 percent to 35 percent is given to streets and roads, and another significant percentage is given to interurban travel. For every car that leaves for work, the area needed for parking equals the footprint of an average house.

Not only do automobiles lead to poor land use in spatial terms; the use of cars in decentralised building layouts is energy inefficient (see Fig. 7). Studies show that a 30-kilometre round trip to work over a year requires as much energy as for heating the traveler's house for the same period. A person who drives more than 10 kilometres a day with a middle-market car consumes as much (primary) energy as he would in the home (Herzog, ed., 1996). Transportation studies have shown that intensification of residential density from 3.9 units/hectare to 39 units/hectare can reduce travel by 40 percent. Studies also show that public transportation becomes viable at residential densities of around 30 to 40 units per hectare. Walking becomes important at around 100 units per hectare (McLaren, 1996). Most studies suggest that the first and most critical factor in reducing car travel is urban density and the mix of uses.

Fig. 7 Energy efficiencies and capacities of common transportation modes (Source: Warren, 1998)

	Approximate passenger miles per gallon (U.S) or equivalent (direct energy consumption)	
	Potential (fully loaded)	Typical (average load)
Urban automobile	64	15
Compact automobile	120	42
Urban diesel bus (for 45 passengers)	215	59
Rail rapid transit	526	42

In aggregate, the adoption of higher density living and working space will tend to reduce the need for car ownership; reduce overall urban travel; reduce the demand for parking space; and increase public transit utilisation. The future objective must be for a more economical use of land achieved through a higher density or intensity in land use in new planning schemes, coupled with a programme of infill developments. This approach will help to cut expenditures for infrastructure and transport, and reduce the exploitation of further areas of available land (e.g. ecosystems or arable land) for development. A social benefit resulting from higher-density residential high-rise development is that it makes pedestrian life at the ground-plane possible. This has the further benefit of reducing the energy, land and pollution demands of the automobile and of other forms of transportation (Van der Ryn and Calthorpe, 1991, p.10). High-density development also reduces per-unit infrastructure costs and thereby improves affordability. It minimises car trips because walking and bicycling become convenient and pleasurable and ultimately the consequence is an increase in the efficiency of public transport (see Fig. 8).

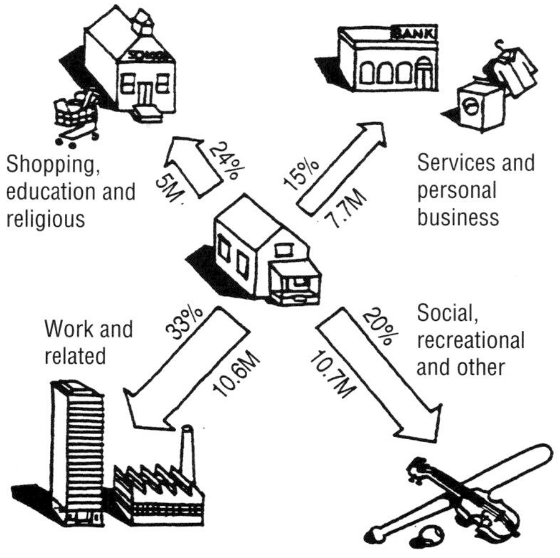
BANK
Shopping, education and religious
24%
5M
15%
7.7M
Services and personal business
Work and related
33%
10.6M
20%
10.7M
Social, recreational and other

Fig. 8 Percent of total household trips each group of activities represents and the average distance travelled for each (Source: Yeang, 1997)

The Smaller Building Footprint

The land conservation justification is that, by building upwards, the land at the ground plane becomes available for recolonisation by nature through natural ecological succession processes. Ecological rehabilitation of land, whether through natural succession processes or replanting, will likely contribute to increasing biodiversity through an increase in vegetated areas. By reducing the land surface covered by buildings and impermeable surfaces (such as roads and paving), the previously devastated land can become unpaved again, which benefits the site's hydrology by improving water absorption and infiltration back into the land as ground water (Katz, D., in Van der Ryn et al., 1991, Warren, R., 1998). The concentration of buildings with a smaller footprint on the land will preserve the areas of land devoted to agriculture and unbuilt vegetated land. Climatically, the concentration and intensification of development into dispersed clusters of centralised development has also the effect of lowering the overall urban ambient temperature of the locality, thereby reducing the overall 'heat island' phenomenon.

The alternatives to the above pattern are widespread low- or medium-rise development. This land use will mean further depletion of bio-productive land; however, it is likely that the further increase in the global average standard of living cannot be sustained unless more land is returned or retained for productive arable use. Furthermore, depletion of the land through low-rise developments will lead to the progressive diminution of natural habitats.

Thus, a disparity has emerged: the decentralised suburban configuration (consisting of low- to medium-rise buildings dispersed throughout the landscape) claims to create habitats that 'coexist with nature'.
But while this perception of the decentralised building style seems ideal and 'green' to many, it is in actuality anti-ecological because the dispersed layout of built forms disrupts the natural ecosystems over a wide land area.

At the same time, decentralised land use creates further problems arising from the provision of new infrastructures requiring more extensive public transportation, expanded accessibility for efficient goods distribution and services, etcetera. In aggregate, providing all these features and services will have greater overall detrimental consequences on the landscape than the concentrated intensive and centralised urban configuration, of which the skyscraper has become an important component.

With this new understanding of the relative total ecological consequences of the two approaches in view, ecological design generally prefers the concentration of construction and the use of small building footprints to reduce current land consumption. Regardless of whether we are for or against the skyscraper, we would still need to come to terms with it as a built form because the skyscraper is here to stay. Urban population projection figures already show that the cities in the world are in fact intensifying; it is projected that in the year 2020, 75 percent of the population of Europe and 55 percent of the population in Asia will be living in cities.

Translated into practice, ecological design means the intensification of existing built-up areas and, where feasible, the use of agriculturally-unproductive land for building. Intensification of urban agriculture (e.g. in roofscape and sky-terrace agriculture) is also called for. Green design also suggests the use of multi-functional spaces, shared spaces and managed spaces in the skyscraper and other intensive building types.

Despite the oil crisis of the early 1970s, which delivered a shock to the high-energy consuming countries of the world, green idealism and architecture have unfortunately not yet gelled into a worldwide common approach by architects. Just as the crisis has not produced any significant pervasive lifestyle change among people, in design there has been no widely affirmed redirection of architecture as yet. Twenty years after the oil shock, a large number of architects are still designing high-energy consuming, multi-storey towers in city centres from Dallas to Beijing.

To reiterate, the imperative for the designers of today to approach large buildings in an ecologically responsible way is clear. However, what should this approach be? There are already a number of architects claiming to have designed ecological high-rises. While many of these designs appear to achieve some level of ecological responsiveness and are 'green' to some degree, the crux of the matter is that these do not fulfill ecological design objectives in their entirety. Instead, what we have are simply some possible solutions out of a number of likely partial solutions contributory to ecological design. The practice of green design is still developing, and in many instances existent ecological design solutions are simplistic. It is important for all to acknowledge honestly that this is so, lest anyone assume that these designs are definitive panaceas and stop seeking further advancements and improvements in subsequent design endeavours. The ecological approach, as outlined in chapter 3, is certainly much more rigorous and complex than is popularly presumed.

For the reasons given above, the ecological design of the skyscraper and other intensive building types needs urgently to be addressed. Otherwise, the consequences might lead to a scenario where our rapidly developing cities in the new millennium will be covered with multitudes of high-energy consuming, high-polluting and high waste-producing skyscrapers and intensive structures. There will likely be immediate benefits in directing our environmental design and research efforts prudently to developing ecologically responsive solutions to large building types. Such endeavours will most certainly show early and significant results, by reducing the level of aggregate negative environmental impacts in the cities where such kinds of buildings are being built as well as by promoting beneficial consequences in the biosphere and on the consumption of non-renewable resources. Green design activities obviously contribute positively to a sustainable future.

The Skyscraper Building Type

What is a skyscraper? We can define the skyscraper as essentially a tall building with a small footprint and small roof area with tall facades. But what differentiates it from low-rise and medium-rise buildings is its special engineering systems, which are required by its significantly greater height. These systems include its structural systems and sub-structural engineering systems; the skyscraper is in effect a vertical 'lever' with the greatest stress at its base. Other special engineering systems required because of its height are its mechanical and electrical (M&E) services systems such as the water supply system, which requires powerful pumps to get the water up to the height of the building from where it is gravity-fed downwards to all the floors. Besides these are the special vertical transportation and movement systems (elevators), special fire-protection devices and other systems. These, in aggregate, differentiate it from other intensive low-rise and medium-rise urban building types such as shopping malls and museums. Special attention needs to be focussed on the ecological design of the skyscraper building type more than on other intensive urban buildings (mostly medium-rise or low-rise), in which many of the well-known ecologically responsive technical solutions (especially solar low-energy solutions) are common.

An argument against the high-rise might be found in the early studies by the Center for Land Use Built Form (LUBFS) under Sir Leslie Martin (in Carolin and Dannatt, 1996). In their research using physical block-form models and theoretical mathematical analyses, the LUBFS group contends that it is possible to build intensive developments using medium-rise courtyard layouts or grouped pavilion layouts instead of tower blocks. I do not dispute this theory here; however, to put this theory into practice, we would need to have land lots of such a size that they could permit these alternative medium-rise built forms to be created (e.g. generally more than 1.5 hectares). But in the real world, we find that architects have to deal with very small land lots (often less than 0.25 hectares) in cities, especially in inner city areas, with many having high plot ratios of 1:4 and above. The design options for such small land lots are limited simply to building upward – the tower built form. Therefore, without other viable physical options, unless amalgamation of the smaller lots into larger lots is enabled, the high-rise tower built form or the skyscraper has become an inevitable form in many of the cities in the world.

Another alternative might be to expand the city boundaries outwards horizontally to accommodate urban growth, or to build satellite cities connected to the existing city by fast light rapid transit (LRT) systems. These two alternatives are certainly viable theoretically, but in terms of land use they also mean further loss of surrounding vegetated land or loss of farmland (agriculturally productive land) which has to be given up for transportation structures,

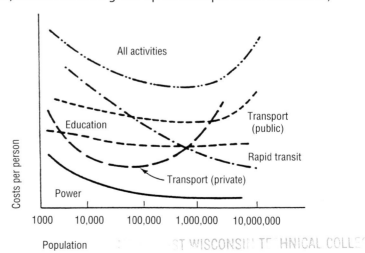

Fig. 9   Threshold to urban density: economies and diseconomies of scale in providing services (Source: Morrill, 1970, in Cadwallader)

infrastructural systems and facilities, as well as for the additional new urban developments. In practice, a balance in land use patterns in relation to the carrying capacity of the surrounding ecosystems to the existing city would need to be arrived at (see Fig. 9).

The skyscraper, then, has evolved as a tall intensive building erected on a small footprint in order to optimise its usually small land area. The Council on Tall Buildings and Urban Habitat (CTBUH) in the United States defines the skyscraper as a tall building that, by virtue of its height, requires its own special engineering systems. The skyscraper's construction is motivated essentially by high land prices and urban transportation expediency. Economically, to enable its developer to get the optimum financial returns from the high urban land prices, the skyscraper has to have sufficient gross floor area (the total built-up space consisting of floors stacked one on top of another) to spread the high land value over the total net floor area (its net rentable or saleable space). In terms of land economy, the intensification of built space over a small footprint (the site area) enables more income-generating floor space to be built over the same small land area. The economic logic of this is that the greater the height of the building, the more built-up space and hence the more the land costs are apportioned upwards, thereby enabling greater financial returns to the entrepreneur (or landowner). For the designer, these realities of building economics dictate that design has to maximise the internal area (net areas) from the permitted buildable gross area for the building on the site to obtain maximum plot-ratios and minimum 'net-to-gross' ratios.

In summary, the skyscraper as a built form is characterised by:
- a small site footprint in comparison to its total built-up space
- tall facades due to its height
- small roof area in comparison to its extensive external-wall area

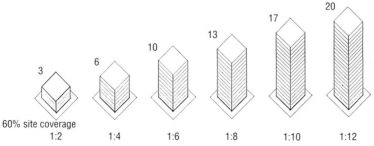

60% site coverage
1:2   1:4   1:6   1:8   1:10   1:12

Note: At a plot ratio of 1:4 at 60% site coverage, the built form is already six storeys excluding carparking. If full carparking is to be provided above ground, the built form is likely 10 storeys or more.

Fig. 10   Land use, built form and plot ratio (Source: Yeang, 1997)

● special engineering systems different from the low-building type, required by its height

In Fig. 10, it is clear that for a site with a plot ratio of 1:4 and with a 60 percent building footprint (i.e. site ground-plane coverage), the resultant built form is already six storeys high, and this is excluding parking provisions above ground. If full car-parking space is to be provided above ground, then the building is likely to be ten storeys or more. Therefore, it is evident that there is no other alternative with such small sites but to go upwards in the form of a tower.

Another aspect of the skyscraper's built form is that it has extensive vertical surface areas (see chapter 6), which affects the extent of solar penetration or build-up into the interior. Formal studies show that it is the circular-plan built form that has the least surface exposure.

It has been stated above that an argument can be made that the skyscraper, by virtue of its intensive use of relatively small plots of land, is ecologically progressive. In fact, there are some who would further contend that the super-tall skyscraper or 'hypertower' (being a building of over 500 metres high) offers even greater opportunities for ecologically-responsive solutions and enables the freeing of land at the ground plane for ecological succession (i.e. return to natural flora and fauna states through natural processes of recolonisation) (Harada, 1996).

It should be made clear that the above discussion assumes that the designer has already carried out the standard built-form and systems studies usual in skyscraper design (e.g. elevator "waiting time and handling capacity" studies, fire-escape route distance calculations, toilet provision calculations and other factors) in addition to pursuing an ecological agenda. In the early stages of the design of the skyscraper, the designer evaluates different alternative floor-plate sizes (i.e. gross-to-net area ratios) and configurations (e.g. wall-to-wall depths) while at the same time keeping in mind the built configuration and massing options of the skyscraper. It is important that at the preliminary stages of design, the designer simultaneously take other considerations into account. These are dealt with in chapters 5 and 6, and include:

- the position of the service cores and how this affects the overall building configuration and layout (in chapter 7)
- the orientation of main facades and window openings (especially in relation to the climatic characteristics of the locality) (in chapter 7)
- the facade design options (i.e. ratio of solid to glass) (in chapter 7)
- the colour of the building envelope (in chapter 7)
- the effects and use of vegetation and planting on the skyscraper's built form (in chapter 7)
- the type of likely building operational systems (in chapter 7)
- the selection of materials and energy sources (in chapter 6) and
- the management of these as potential waste products (in chapters 6 and 7).

The above are in addition to the totality of ecological considerations discussed in chapter 3.

In summary, the ecological design of the skyscraper and other large buildings as the high-energy and materials-intensive urban building types of cities today is a matter deserving urgent attention from our ecological designers. This development is vital because of the skyscraper's ubiquitousness worldwide.

While some might contend that this building type might not continue to be built in significant numbers in the West (i.e. developed countries), there are a great many of these tall buildings already in existence (for example in New York City, Houston and Chicago) which may need to be reconditioned and adapted to meet acceptable standards of ecological design. To abandon these wholesale now would create other problems such as the production of waste, besides resulting in attendant planning problems.

Looking forward to our discussion of green design, we might redefine the skyscraper and other intensive building types from the viewpoint of the ecologist as being simply a high volume of inorganic mass that is brought together by the designer, then assembled and concentrated onto a small footprint site, and which operationally consumes large amounts of non-renewable energy resources, emits large amounts of wastes (mostly paper products and waste heat) and affects disproportionately the energy flows of the natural ecosystems of the locality. At the same time, its users (being only one of the multitude of biotic components in this inorganic mass) consume further amounts of non-renewable energy resources in coming to and departing from the building by various means, usually travelling over impervious road surfaces that detrimentally reduce rainfall infiltration back into the groundwater. These constitute in effect, some of the key aspects of the skyscraper's architecture that need to be redressed by ecological design.

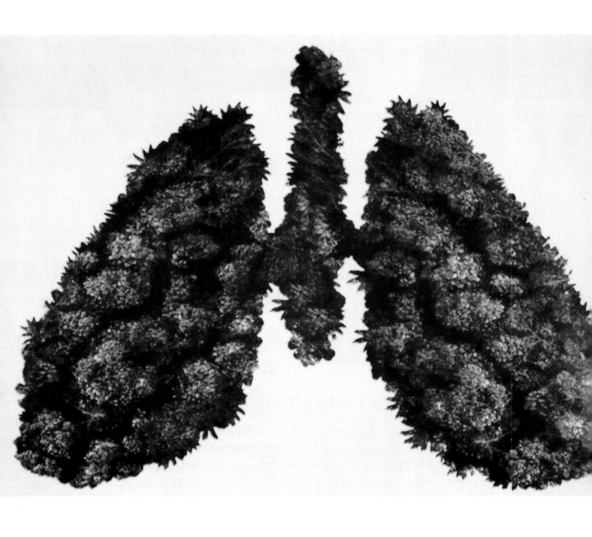

Chapter Two

What is
Ecological Design?

What is Ecological or 'Green' Design?

If we are to regard ecological design as a positive rather than a negative endeavour and as designing to meet ecological objectives rather than merely to cope with ecological constraints, then our first task is to define ecological design, and then to identify corresponding objectives.

Before doing so, we should first ask whether our architects and designers are at present theoretically and technically equipped at all to respond to the new demands that arise from green design. The answer is likely no (MacKenzie, 1997, Papanek, V., 1995, p. 48), for almost all architects (except landscape architects) today have been trained without any serious background in ecology and environmental biology. Hence it is contended that ecological design calls for a rapid and fundamental reorientation of our thinking and design approach with regard to the creation of our built environment.

Connectedness: Architecture as Applied Ecology

To design in an ecologically-responsive way will require a fundamentally different view of our relation to and our place within the natural world; it will require a departure from the limitations of current science and the social, political and economic context which implicitly valorizes human enterprise as dominant over and essentially independent of nature. Ecological design requires the architect to regard and to understand the environment as a functioning natural system and to recognise the dependence of the built environment on it. This sense of the interdependence of the constructed and the given (i.e., 'natural') environments could be called 'connectedness.' Before we can proceed with our strategy of green design for our intensive building type, we need not only to first define and understand what constitutes green design, but also to understand its premises, for it would be counter-productive for the designer to leap into green design without understanding and agreeing to such basic principles as connectedness.

Central to ecological design is of course the concept of "ecosystem" itself, which requires an analytical understanding of the environment – and, specifically, the particular site in question – as consisting of biotic components and abiotic components acting as a whole (see Fig. 11). This is crucial to our ecological approach. For instance, a mechanistic use of computer software for analysing energy conservation, airflow and temperature acoustic factors which does not take into consideration the biological components (e.g. the flora and fauna) or the edaphic factors of the place could hardly be called ecological design at all. Similarly, if a design approach does not take into account the holistic aspects of the environment, it is certainly not ecological.

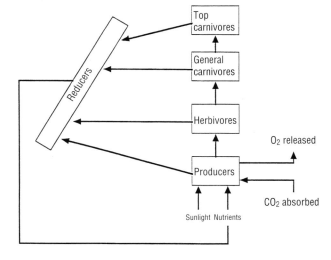

Fig. 11   The cycling of materials within an ecosystem (Source: Boughey, 1971)

Simply stated, in ecological design, we need to evaluate the consequences of:
- if we build (see chapter 4),
- where we build (see chapter 5),
- what we build (see chapters 4 and 6) and
- how we build (see chapters 6 and 7).

The practice of ecological design is essentially 'applied ecology' or the practical application of ecology to human intrusion into the natural environment (in which building is simply one of a multitude of man's activities that affect the environment).

A prerequisite then for ecological design is an understanding of the basic systemic concepts of ecology and their applications. This is necessary to enable the designer to see how his endeavours, as human interventions in the environment (whether in agriculture, building development, the building of roads, and so forth) can be carried out in such a way as to integrate with the natural systems (e.g., with minimum disruption of the ecosystem, with prudent use of the earth's non-renewable resources, and with the activities associated with the designed system symbiotically compatible with the processes of the ecosystems). Meeting these objectives is crucial in the ecological approach.

Sustainable Development

The continued degradation of the biosphere through over-exploitation and abuse not only diminishes its ability to produce essential resources but also its ability to recover from such abuses. A prerequisite for sustainability is the maintenance of the functional integrity of the ecosphere so that it can remain resilient to human-induced stresses, as well as biologically productive. Non-renewable resources, as finite assets, must be used or transformed in such a manner that they remain useful and accessible to future generations. Seen in this light, the basis of the concept of 'ecological design' is not that it is a retreating (nor a losing) battle, constantly seeking to minimise impacts on the natural environment and retard degradation. Rather, ecological design can be seen as environmentally beneficial and productive, a positive contribution to the natural environment. Further, ecological design should be a positive act of repair, restoration and renewal of the natural systems of the environment (Berkebile, R., in Zeiher, 1996, p. 31). I contend that green architecture as sustainable architecture is designing with nature in an environmentally responsible way as well as a positively-contributive way. Achieving these two objectives simultaneously by design is probably the greatest challenge confronting the ecological designer today.

All design endeavours in relation to the earth's ecological systems of course refer to the future; they therefore can and should be prognostic and anticipatory. For example, buildings should be designed with prior regard for the recoverability, re-use and recyclability of their constituent materials and components. This is exemplified in the concept of sustainability, which is described as 'meeting the needs of the present without compromising the ability of future generations to meet their own needs' (Brundtland, 1987).

This makes the concept of sustainability a complex one and therefore involves both subjective as well as objective (e.g. quantitative) decisions affecting human welfare both in the present and in the future. More specifically, ecological design involves literally thousands of ways in which a built system and its users connect to the natural world.

Ecology is about linkages, interdependence and creative adaptation as opposed to compartmentalised causality. Ecological design therefore can be seen as a holistic connection, entailing the prudent management of energy and materials in the built system (see chapter 4) alongside the ecosystems in the biosphere; it will include both those design endeavours that reduce the detrimental impacts of this management on the ecosystem and those that try to integrate positively with the natural environment. Furthermore, the meeting of these objectives is not a once-only occasion, but has to be managed and monitored over the entire life cycle of the built system, i.e. from its source to sink (in Yeang, 1995). This complexity is dynamic, extending over time; this topic is discussed further in chapter 3.

The issue of sustainable development (of which ecological design is an element) at the global level is now beginning to be seriously addressed by most governments in the world, as well as by intergovernmental agencies and fora. At the personal scale, concern for the environment has led some to seek 'green' alternative life-styles (Slessor, 1997).

At the level of the professional designer, what might be regarded as part of a slow but gradual greening of architecture has already engendered some results, such as the establishment of more stringent thermal performance standards in buildings (e.g. BREEM in the UK), the eco-labelling of building materials and products (particularly in Germany and Canada), the intention by some designers to green the design and building process, and the increasing monitoring of the energy performance of buildings in use (by many of the architects and engineers in Europe and in USA) and a greater awareness of ecological factors on-site and the importance of biodiversity.

Ecological design in most instances might be better regarded as an act of rapid-prototyping; such solutions are the best that can be achieved now, but with consideration of the ecological component and while acknowledging that subsequent improvements are essential in the next prototype.

In ecological design, we need to acknowledge that many of the earth's ecological systems and processes are simply too complex to be quantified and represented in their totality. Nevertheless, ecological design, as I shall show, remains a complex proposition and involves the resolution of a large number of considerations of multiple sets of interactions (or corrections) (see chapter 3). Architects, designers, engineers and all those whose work affects the environment must somehow make everyday design decisions. They need to take decisive action on issues daily on the basis of the environmental information that is available at the time. It is thus vital that the present inadequate state of environmental knowledge not be used as a justification for the evasion of the ecological approach (including preventive or corrective action) and the evasion of responsibility for the environmental impact of building projects.

The significance of taking design action based on a proper and full understanding of ecological criteria is obvious, because the design and planning decisions that are made in the present will not only have an immediate effect on human society and the environment, but also could influence environmental quality for subsequent generations, thereby contributing, to a greater or lesser extent, to a sustainable future. In the process of ecological design, we should proceed on the basis of what is already known in an anticipatory way, rather than with ignorance or, at worst, by excluding environmental considerations in their entirety. Adopting a design approach that is deemed the best that can be ecologically achieved today (erring on the side of caution) may in many instances be better than waiting sometime tomorrow for the perfect, comprehensive solution – by which time extensive devastation of the ecosystems might already have taken place.

At the same time, if unavoidable design decisions need to be made, we must also be aware of the hazards of a piecemeal approach to ecological design, which would not effectively address the global issue of environmental degradation. Thus, any urgent steps needed to stem the continuation of the destruction of the environment by high energy-consuming and high waste-producing urban buildings must be made in full regard of the consequences on the environment (see chapter 3). Therefore, our efforts toward the green design of the intensive building type, although imperfect and in some instances failing to fulfill the totality of the criteria of a comprehensive ecological approach (due to current technologies' deficiency and other factors) must at the same time be anticipatory in eliminating negative impacts as much as possible. Our design must contribute to the greatest extent possible to reducing the overall environmental impact of such intensive buildings on the environment while allowing for future enhancements, improvements, and replacements.

In the long term, it must be acknowledged that at the global and national level, changes in the economic, social and political systems based on holistic ecological principles are crucial if the objectives of a sustainable future for mankind are to be met. Although these lie beyond the realm of this book as well as beyond the sphere of influence of the designer, the fundamental principles of green design as applied ecology remain relevant to the development of global systems of green politics, economics, physical planning and social systems.

The Basis for Ecological Design

Besides understanding the principles of the ecosystem concept, it is crucial for the designer to understand some of the fundamental premises of the design approach. The objectives for ecological design are as follows:

● Ecological design acknowledges the resilience of the natural environment and its limit.

The designer's built system and its operations make use of the resilience of the natural environment (e.g. its processes and renewable resources) to absorb some of their negative impacts. Despite the inherent resilience of the natural systems in the biosphere, the designer must be aware that there are limits to this resilience and to the natural environment's carrying capacity. The extent of this resilience and capacity vary depending on the locality.

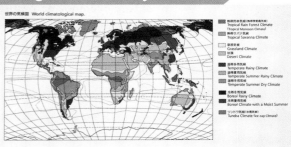

A fundamental aspect of natural systems is ensuring that ecosystems and the entire biosphere are relatively stable and resilient. The ability to withstand disturbances and to recover from regular 'shocks' is essential to keeping the biosphere's life-support system operating. Maintaining the integrity of the web of species, functions, and processes within an ecosystem and the webs that connect different systems is critical for ensuring stability and resilience. As ecosystems become simplified and their webs become disconnected, they become more fragile and vulnerable to catastrophic, irreversible decline. In manmade changes such as global climate change and the breakdown of the ozone layer, the biodiversity deficit, the collapse of fisheries, frequent outbreaks of red tides, and increasingly severe floods and droughts, there is now ample evidence that the biosphere is becoming less resilient.

Some of the benefits (or 'outputs') provided by natural systems include:

- raw materials production
- pollination
- biological control of pests and diseases
- habitat and refuge protection
- water supply and regulation
- waste recycling and pollution control
- nutrient cycling
- soil formation and protection
- soil building and maintenance
- ecosystem disturbance regulation
- climate regulation
- atmospheric regulation (CO_2 absorption and O_2 release)

Many such natural benefits arise from the environment's ability to regulate and recycle water, nutrients, and waste. One of the most basic aspects of the cycling and recycling system is that water falls as precipitation, running across the landscape to streams and rivers and ultimately to the sea. But human disruptions have impaired this ability to filter and regulate water, to recharge ground-water supplies, and to move nutrients and sediments – indeed, to support life.

Human actions have even changed the fundamental forces of nature by removing natural plant cover, ploughing fields, draining wetlands, separating rivers from their flood plains, and paving over land. Continued and extensive urbanisation and land use have

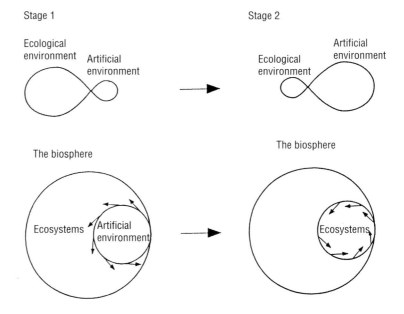

Fig. 12 Biosphere saturation (Source: Yeang, 1997)

changed the relationship of the manmade and natural environments – indeed, they have reversed them, so that the human sphere
has gone from being the 'contained system' within nature and the
earth to being the 'containing one'. Ecosystems, increasingly saturated with manmade (synthetic) systems (see Fig. 12) (after Chermayeff and Tzonis, 1971), have been losing their ability to self-regulate
and to assimilate human outputs. A further effect has been the loss
of biodiversity. These trends have regional and global dimensions
(see chapter 5).

● Ecological design acknowledges the importance of biodiversity.

Where urban development has reduced the number of species and
the size and integrity of ecosystems, the consequence has been a
reduction in nature's capacity to evolve and create new life. In just
a few centuries, we have gone from living off nature's "interests" to
spending down the "capital" that has accumulated over millions of
years of evolution, as well as diminishing the capacity of nature to
create new capital.

More specifically, building activity (e.g. site clearance, construction works, etc.) usually results in some degree of ecosystem simplification – that is, from a diversified state to a less complex biological state. The consequence of such simplified systems is a lack of
that resilience which allows them to survive short-term adversities
or long-term alterations such as climate change. Often this results
in reducing the flexibility of the relationship between the manmade environment and the ecosystem, while simultaneously
increasing the ecosystem's constraints on the manmade environment. The overall effect is that humanity and its built systems have
become not less dependent upon the functioning of the ecosystems
within the biosphere, but on the contrary, have now become more
dependent. Diminished ecological capacity means that the scope
for human action is also narrowed and as humanity's options
reduce, so must we proceed with greater care in the use of the
natural environment (see chapter 5), particularly with regard to
maintaining biodiversity.

● Ecological design has to take into account the connectivity of ecological systems (see Fig. 13).

As we have seen, a designed system of any kind must necessarily
interact with its environment. It thereby interacts with and affects
the earth's ecology to a greater or lesser extent. But owing to the
overall loss of resilience in the ecosystem mentioned above, over

Fig. 13 The three environmental zones of air, water and land and their interaction with the biotic factors (Source: Yeang, 1997)

Land   Air   Water   Others

Some environmentalists wrongly conceive the environment in terms of discrete environmental zones which do not interact with one another

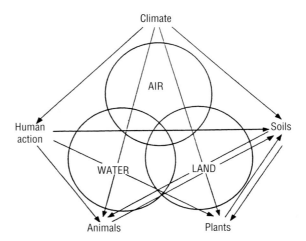

time the ecosystem grows more constrained, with the result that the ecological element of the designed system grows correspondingly more important. Therefore, greater monitoring will be required. It is the components (organisms, populations, species, habitats, etc.), processes (nutrient cycling, carbon cycles, ecological succession, etc.) and properties (resilience, health, integrity, etc.) that make up the ecosystems that provide us with a life-supporting environment.

⚫ Ecological design must acknowledge that manmade synthetic ecological systems can never adequately duplicate the complexity of natural ecological systems.

In contradistinction to the model being developed here, there are those, among them designers, who have held – erroneously – that the absorption capacity of global ecosystems is robust enough to withstand the impacts to which we have subjected it. They bolster this contention with the claim that, even as human beings intervene in and interfere with the environment, it will be possible to sustain ecosystems by devising artificial 'subsystems', whose role is to replace the ones provided by nature. This view holds out the hope that technology will allow human beings to stay ahead of nature, in effect, until such point as technology renders us completely independent of nature (e.g. Landers, 1966). This mechanistic and utopian model, based on a blind faith that technology will be able to surmount all environmental problems in time, has already been applied in design practice. A prominent example is the development of a type of design incorporated into the built environment's life-support systems and known as 'artificial controls' or 'controlled environments'. In such a built system, all the existing natural self-regulating mechanisms of the environment are substituted by manmade mechanisms which control the structure's mechanical and electrical systems, heat, air conditioning, lighting, waste management systems and other structures. This form of asystemic control has significant drawbacks, primary among them the fact that the interaction between the manmade system (the built environment) and the existing self-regulating systems of nature becomes concentrated in human hands (Goldsmith, 1971); and because manmade systems are only gross approximations – or rather, simplifications – of natural systems in all their complexity, these 'artificial controls' are prone to failure. Indeed, it is a fantasy that such artificial systems can ever entirely take the place of self-regulating natural systems.

● Ecological design must seek to repair and restore ecosystems.

Our designed systems must aim to repair and restore devastated natural environments; properly restored or constructed ecosystems can provide many of the services people require. For example, swamps and wetlands were once viewed as wasted land, productive only if drained or filled. Today, their roles in cleansing water, recycling nutrients, recharging aquifers, controlling floods, and supporting productive resources such as fish, wildlife, and wild produce are being recognised. Their function as a line of defense in protecting coastal and ocean ecosystems from land-based pollution is clear, as is their ability to protect coasts from storms. Many countries are now using the assimilative capacity of natural or created wetlands as a cost-effective way to control and filter storm water and industrial and agricultural runoff, to decompose human waste, and to cleanse water.

Another example is with agricultural lands, where buffer strips of trees, together with restored or constructed wetlands, can reduce runoff of major pollutants such as sediments, phosphorus, and nitrogen by 80–100 percent. In both industrial and developing countries, wetlands are a low-cost, low-technology alternative to sewage treatment plants.

Restored ecosystems can also bring back some of the flood control capacity that nature once provided – and at a lower cost and with greater effectiveness than structural alternatives such as dams and levees. Studies have found that with each 1 percent increase in wetlands, flooding downstream is decreased by 2–4 percent; watersheds that are 5–10 percent wetlands can reduce the peak flood period by 50 percent compared with watersheds that have no wetlands.

Even more effective than mitigating the impacts of activities or rejuvenating degraded ecosystems is maintaining healthy ecosystems. A few studies have measured the aggregate values of ecosystems and the unanticipated losses of economic, social, and ecological benefits that can ensue when the systems are degraded.

As the resilience and the assimilative ability of the ecosystems in the biosphere continue to be reduced, an eventual limit to the extent to which external manmade controls may be permitted to replace the complex, ecologically self-regulating ones will be reached. The designer must appreciate that although an ecosystem is able to assimilate a certain amount of impairment to its processes, it has a definite limit to its assimilative ability, and that if an ecosystem is not to be permanently impaired, the designer must ensure that all subsequent actions and activities that take place in it must remain subject to the limitations inherent in the ecosystem and its components. In most instances, these limitations can become apparent to the designer only after a proper examination of the ecosystem of the project site and its properties has been undertaken (see chapter 5). Therefore, before any building can be erected on a site, its ecology must be mapped and studied. Ecological design must seek to minimise its dependency on the resilience capabilities of the natural environment both locally and globally. Where this cannot be avoided, it must ensure that this resilience is not pushed beyond its limits.

● Ecological design seeks a symbiosis between manmade systems and natural systems.

Nature has built-in ecological controls, such as the nurturing of innumerable species, that are not harvested directly but which provide important 'free' services. Thus, species pollinate crops, keep potentially harmful organisms in check, build and maintain soils, and decompose dead matter so it can be used to build new life. Nature's 'service providers' – the birds and bees, insects, worms, and microorganisms – show how small and seemingly insignificant things can have disproportionate value. Unfortunately, their services are in increasingly short supply because habitat fragmentation and destruction have drastically reduced their numbers and their ability to function. In the present state of the built environment, people have created a situation where they must now return to the natural ecological controls, develop new ones, or design some new combination.

At present, it seems unlikely that people can construct adequate artificial control systems out of engineering hardware only, while completely ignoring the natural climatic and ecological systems. Therefore, the 'green,' sustainable design option is to integrate the manmade systems with those of the environment symbiotically, so as to make use of existing natural controls (e.g., passive systems), and/or combining both manmade and ecosystem (or biological) control structures.

In the case of the intensive building type such as the skyscraper, the discharges need to be recycled or reused within the built system itself or within the larger overall urban context of the city as much as possible (e.g., through recycling or reuse of wastepaper, office and residential products, rainwater and wastewater, waste heat, etc.). If we are unable to recycle these wastes within the built system, then they should be recycled within the entire urban system in which the designed system is located (i.e., within the city's infrastructure). Essentially what this means is that at the onset of our design process, the larger-scale systems of urban recycling, reuse and repair must be taken into account. This applies not just to volumetric waste but also to the molecular waste (e.g. CO_2 discharges from the building's stand-by generators, heating systems, etc.), as well as emissions of waste heat (e.g., thermal discharges, etc.) (see chapter 4).

The current state of site analysis by designers does not address these issues. At present, a designer generally looks only at the physical features of the site which the proposed designed system will occupy. This form of site analysis gives the designer a basis for determining the best location for the structure(s), creating the layout, making provision for vehicular access, and other aspects of the design, including height, shape, etc. Ecological design, however, goes beyond these physical features of the site to include biological criteria and a knowledge of the surrounding ecosystem and its processes. The designer will have to understand and master these aspects of the proposed site to determine what type of building could be allowed in that ecosystem by the criteria of sustainable development and 'green' design principles. To create a symbiosis between the designed (manmade) and natural systems is the designer's task; on a physical level, the building's systemic features, processes and functioning have to be integrated with the ecosystem's to avoid undesirable or destructive impacts as a result of human intervention in the natural environment (see chapter 4).

● Ecological design takes into account entropy in natural systems.

The designer needs to be aware of entropy in natural systems in the biosphere and not contribute by his designed system to further accelerating the entropic processes in the natural environment.

Entropy can be imagined as the degree to which the universe 'runs down' over time. Entropy increases in every natural process (Berry, 1972), and can be seen variously as representing disorder or dilution within a system. For our purposes as designers, one could also define entropy in terms of the amount of dissipation from a particular system or as the energy that allows the system to do work; the resulting dissipation could either be internal or could be expelled from the system to its environment (Walmsley, 1972). Entropy can also be conceived as the degree of dilution or disorder in a system.

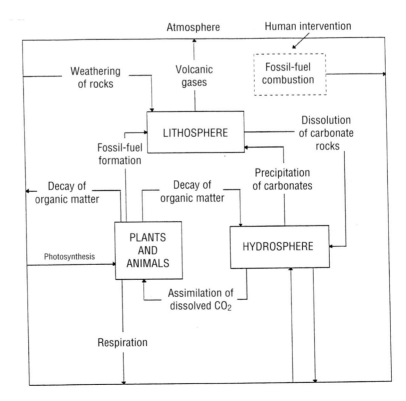

Fig. 14 Disruption to the carbon cycle in the biosphere by human intervention through fossil-fuel combustion (Source: Yeang, 1995)

Human activity throughout history has expropriated natural resources from the earth; as discussed above, the strain on ecosystems due to human activity has eroded much of the biosphere's natural self-regulating ability (its ability to 'heal') – with the result that entropy has accelerated and increased. Perhaps the clearest example of human interference in a natural cycle or system relates to the carbon cycle. Fossil fuels are of course a key energy source, and are also non-renewable. Figure 14 illustrates the cycle of the biosphere responsible for their production. The high consumption of fossil fuels by humans is in essence a short-circuiting of the biogeochemical cycles. Human intervention in the form of resource use accelerates one aspect of the cycle (combustion or consumption) to the point where it exceeds the system's ability to regenerate naturally. A similar short-circuiting obtains in the case of other elements and resources, such as iron and other metals, nitrogen, phosphorus, and mercury (Bowen, 1972; Holdren and Ehrlich, 1974). It goes without saying that these materials have been used intensively in building and construction, and this consumption has outstripped the capacity of the natural cycles responsible for their formation in the environment.

As though to defend the built environment from its implication in this process of entropy increase, some designers have mistakenly used the orderly nature of architecture to claim that design is anentropic. In this view, design and building counteract the effects of entropy in the biosphere, thus bringing 'order out of chaos'. But this is an oversimplification of what architecture does (after Bertalanffy, 1968). Entropy necessarily increases in an open system. A living organism is itself an open system but it maintains itself in a steady state by importing energy-rich materials, consuming the energy in complex organic molecules and then expelling the simpler end products of the process into its environment. It appears, then, to avoid an increase in entropy. But if we view the environment and the system holistically (and thus, incidentally, in line with the 'green' perspective of our ecological design method), then the total energy exchange taking place in the system organism-plus-environment still conforms to the second law of thermodynamics. Indeed, this will always be the case, for only when the living system is seen in isolation does one have the impression that it is tending toward a higher state of order and complexity in an entropy-free way. Actually, the increasing order and differentiation of the open system (the organism) is at the expense of energy won by oxidation and other energy-yielding processes, and some of its internal reactions always produce a degree of entropy. Processes like growth, decomposition and death are facets of a slow exchange taking place within a steady state; each is accompanied by the expenditure of energy. The second law of thermodynamics is not violated when the combination of the open system and the environment in which it exists is seen in its entirety, and entropy increase still takes place. This inescapability of the entropic process only underscores the importance of a holistic view and knowledge of ecosystem properties in the ecological design method.

● Ecological design acknowledges that the environment is the final context for all design.

The designer must expand his or her previously restricted concept of the environment (when considering a project site) to incorporate the ecologist's more holistic concept of the environment (see chapter 5). Ecologists contend that the environment of any built system must be seen in the overall context of the ecosystem in which the built system is located, and in turn that this ecosystem unit also exists within the context of other ecosystems on the earth. When the term 'environment' is used in the present work, it refers to this totality.

In ecological design, the project site for our building must at the outset be conceived as part of the locality's ecosystem by the designer and as an environmental unit consisting of both its biotic and abiotic (living and non-living) components functioning together as a whole to form an ecosystem. Before any human action can be inflicted on the project site, the site's ecosystem features and interactions must be identified and fully understood (see chapter 5). The living, natural world must be regarded as the matrix for all design, and therefore our design should follow, rather than oppose, the laws of life in the ecosystems (e.g. Todd, N.J., and Todd, J., 1994).

Therefore, the designer cannot view all project sites as uniform, or as economic commodities with uniform ecosystem features. In the same way that no two biological specimens are exactly the same, each location for building is ecologically heterogeneous even though some superficial similarities may appear. A site's ecosystem has physical attributes, organisms, inorganic components, and interactions (see Fig. 15) which are unique. Site analysis has to begin by capturing these unique features; only on that basis can the designer take decisions regarding their use, preservation, or conservation. Ecological design therefore requires that an analysis of the ecosystem of the project site, its components and its carrying capacities be made before any construction activity can be permitted to start (see chapter 5).

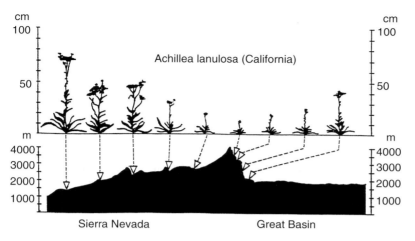

Fig. 15   Genetically identical plants develop differently according to the environment (Source: Yeang)

The uniqueness of the ecosystem features of a site is mirrored by the specificity of designed systems. A built system that works for one particular site may not be translatable to another, even if there are superficial similarities between them (see chapter 5). The site of a proposed building must be evaluated in terms of its ecological components even if it appears devoid of any ecological features; even ground-water conditions, topsoil, existing trees, and other such elements may have ecological consequences.

● Ecological design acknowledges that the built environment is dependent upon the earth as the supplier of energy and material resources.

In addition to the built environment being dependent on the natural environment for its ecosystem processes (see above), the designer has to be aware that the physical substance and form of the built environment are constructed from the renewable and non-renewable energy and material resources which are derived from the earth's mantle and its ambient environment (see Fig. 16). However, in addition to the built system's dependence upon the earth's ecosystems, the built system is also dependent upon the earth for its continued existence in its operations, for the earth is a supplier of energy and material resources (see chapter 6). As a consequence, ecological design is a form of prudent management of the use of these resources in their life cycle (see chapter 4).

The earth is essentially a closed materials system with a finite mass, and all the ecosystems within it, along with all of the earth's material and fossil energy resources, form the final contextual limit to all our design activities. All design inevitably takes place within the confines of this limit. For example, one of the planet's first eco-system functions was the production of oxygen over billions of years of photosynthetic activity, which allowed oxygen-breathing organisms such as ourselves to exist. Our future existence will continue to depend on ecosystems that are responsible for maintaining the proper balance of atmospheric gases such as oxygen and carbon dioxide. There is no technological substitute for this vital service which nature provides.

Humans have begun to impair this system by generating too much carbon dioxide and other greenhouse gases, and by reducing the ability of ecosystems to absorb carbon dioxide. The benefits of intact forests for global carbon sequestration alone are self-evident. The consequences of disruption of this natural function are beginning to be evident in the form of global climate change. Maintaining nature's ability to regulate local and global climates will be even more valuable under the predicted climate change scenario. At the onset of ecological design, the designer has to acknowledge this limitation.

Ecological design demands a rational use of the ecosystem's processes and non-renewable resources. Such was not the case in the past, when designers imagined that the environment was essentially infinite – both in its capacity to supply resources and in its ability to act as a sink or dump for the discharges of waste materials. Such a view, as is obvious from the foregoing discussion, can

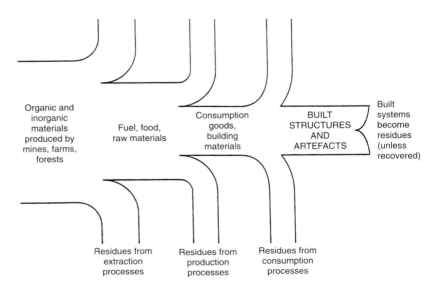

| Organic and inorganic materials produced by mines, farms, forests | Fuel, food, raw materials | Consumption goods, building materials | BUILT STRUCTURES AND ARTEFACTS | Built systems become residues (unless recovered) |

Residues from extraction processes

Residues from production processes

Residues from consumption processes

Fig. 16   The built environment as part of the flow of energy and materials (Source: Yeang, 1995)

Residues: solids, particulates, gases, heat, liquids, etc.

no longer be maintained, and the ecologically minded designer must keep in mind the very real limitations of the biosphere, and the finite capacity of an ecosystem to recover from the loss of resources (appropriated for human development) on the one hand and the influx of waste products on the other.

An ecological approach to design, therefore, is a conservation-minded one. The question of non-renewable resources is critical. The production, operation, and eventual disposal of the designer's system (the building itself) will consume a certain quantity of resources, of which the designer must be aware. The designer of the built environment must also be informed about the degree to which resources are utilised or reused – in essence, the efficiency of resource consumption by the built system. One factor is the spatial accommodation that he or she has designed into the system, which may be in excess in terms of the building's requirements (see chapter 4). Should there be a difference in this provision of accommodation, it will reflect the efficiency of energy and resource use by the building in question. Such differences can also be quantified as a measure of the built environment's impact on the biosphere and its consumption of earth resources.

Ecological design in effect is design which:
● utilises renewable resources ideally at rates less than the natural rate at which they regenerate;
● optimises the efficiency with which non-renewable resources are used.

Ecosystems are the end point of waste, discharges, and all other outputs of human built systems; this is the idea of the biosphere as 'sink'. But, because ecosystems are finite and their ability to absorb these outputs is likewise finite, limits have to be placed on the discharge of waste products from the building lest the surrounding ecosystem's assimilative systems be overwhelmed. In a larger view, we see that the 'life' of the entire built system is finite as well, and hence the designer has to consider what will be the final fate of the components of his building. When its usefulness is at an end, the building itself becomes waste, and its materials have either to be recycled or disposed of. Thus we can distinguish two aspects of the waste equation confronted by the designer, that relating to the amount of waste that will have to be processed during the building's life, and that which will have to be dealt with when, at the end of that life, the structure itself has to be disposed of. It is the ethical and professional responsibility of the ecological designer to consider both of these elements, for the responsibility for a building does not end upon handover to the owners at the completion of construction; the designer's responsibility has to be from source to sink, covering the whole flow of the built system's components during its life cycle.

● Ecological design acknowledges that all design has a global impact because of ecosystem connectivity.

The designer must not only see the impacts of his designed system in the restricted confines of the ecosystem in which it happens to be located. Because ecosystems are interrelated, the effects of the built system are passed through one ecosystem to another and may ultimately be global. As described above, the ecosystem in which the building exists is itself made up of systems, cycles, and functions which interact with each other; in a similar way, but on a larger scale, ecosystems interact with other ecosystems, generating further effects on the biosphere as a whole (in Todd, N.J., and Todd, J., 1994, et al.).

A further point is that environmental contamination does not respect manmade borders. However, a building site – at least in the current understanding of the architectural and legal professions – is usually described by its lot boundaries in a legally recognized way, much as a country is defined by internationally agreed borders. An ecosystem, by contrast, has evolved within natural boundaries, which may be criss-crossed by human lines and divisions, which, of course, the ecosystem does not respect. Just because the environment of the designer's site also has several or many other building lots contained within it does not mean that the designer can think of his or her particular site as discrete, an isolated entity which exists only within the legally specified, manmade lot lines. A site is not a square on a map, but in the world; actions taken on that site have effects on local ecology that will extend to the other human-made parcels (i.e., building lots) in the same environment and beyond. As we have seen, the scale of such impacts of a design can be local, regional, continental, and biospherical all at the same time.

In the case of the urban building such as the skyscraper, most such sites, being urban, would usually have been already extensively degraded and rendered devoid of any biotic components. In such instances, the remaining environmental effects that the built system would have to take into account would be impact on the local micro-climate level (e.g. air pollution, thermal emissions etc.), impacts on surrounding buildings, and emissions and discharges of waste from the built system into the city's infrastructural system which are then discharged elsewhere into the environment, both locally and globally (because of the connectivity of all natural systems in the biosphere).

● Ecological design involves the management of outputs from the built environment into the ecosystem.

The designer must be aware that all designed systems, as open systems, emit outputs. These outputs, as waste, enter the surrounding ecosystem, whether they are in solid, liquid, or gaseous form. In some cases, the outputs are brought back into the built system and recycled and reused, while other wastes may unavoidably be discharged into the ecosystem and have to be absorbed by the environment. As has been mentioned before, the built system depends on its surrounding natural environment for the assimilation of the waste it produces. Therefore, the designer cannot simply take the view that, once the wastes exit the built system, they no longer have to be thought about, as if they had somehow disappeared by passing the boundary of the built environment. Outputs from urban buildings such as skyscrapers, as from any built system, have to be absorbed into the ambient world; this may or may not require some degree of treatment to facilitate their assimilation by the ecosystem. Determining whether and what level of pretreatment is necessary must, of course, be the responsibility of the designer, who will have to take into consideration various limiting factors. Meteorological features of the local environment determine the rate of waste dispersion by air; rainfall and the rate of groundwater run-off set limits on how much waste can be tolerated by riverine and other water systems; soil conditions affect not only land-based waste disposal, but because the environment's systems are linked, they will determine waste-water reclamation and other factors. Similarly, topography, like soil conditions, affects the possibility of waste disposal through landfills, but also plays a part in flooding and erosion – environmental dangers that the designer must also take into account.

This complex panoply of local environmental factors will determine the resilience of the ecosystem in which the building is situated. When outputs of waste exceed the capacity of the local environment to assimilate them, ecological damage results.

The designer must anticipate and avoid such a scenario by keeping outputs below a threshold determined by the ecosystem's assimilative capacities. Hence, the designer has to have a complete knowledge not only of the prospective waste outputs of the structure he or she is designing, but also a detailed understanding of the ambient ecosystem down to the level of each biological, chemical and physical cycle. This background of knowledge amounts to an ecological profile of the locality in which the built environment is to be placed; it goes without saying that this eco-profile has to be compiled in advance, before any human action is taken that impinges on the environment.

Design must include the estimation (before construction) and monitoring (after construction) of the outputs from the built system. As stated earlier, ecosystems can tolerate a certain amount of human intervention, but there is a limit beyond which an ecosystem becomes irreparably damaged. The outputs from the designed system must find a place somewhere in the biosphere, whether within the built environment ("in-use") or as waste to be voided elsewhere.

● Ecological principles require all design to be regarded in the context of its physical life cycle.

Early in the design process – i.e., the preliminary phase – the designer must have extrapolated to the degree possible all the effects of the building over its entire projected lifespan. These projections obviously include estimated impact on the environment; by a process of feedback, the estimates then are factored into the design itself before any building has taken place (indeed, before the design has been finalized), so that the design of the built system anticipates its own impact on the ecosystem. As has been mentioned, the designer has a source-to-sink responsibility encompassing the designed system's total requirements and outputs of materials and energy from construction through the period of its useful life and beyond, to its eventual disposal. It is also important to visualize the flow of resources through the building, from their initial extraction from the earth to their return to the environment as waste products of the built system.

This is particularly important because the interactions between ecosystems are dynamic processes and change over time. Ideally, the designer should anticipate the impact and the performance of the designed system in the local ecosystem throughout the entire span of the designed system's life – during which the states of the ecosystems do not remain static but are themselves changing. In architectural practice, the current restricted range of responsibilities of the designer would need to be expanded to include the responsibility for the environmental impact of the designed system over its whole useful life. Simultaneously, some form of environmental monitoring would be needed to check the impact of the designed system on its environment during this period, including the changing state and response of the environment.

● Ecological design acknowledges that all building activity involves ecosystem spatial displacement, and some displacement of energy and materials.

Even a well-designed building, merely by its existence, spatially displaces the environment around it. Its physical presence on the site changes the ecosystem's composition. The siting, layout, structure, and component systems of a building are all aspects of this physical presence, and therefore have to be evaluated in terms of their effect on the corresponding systems embodied in the environment, including spatial pattern and functioning.

● Ecological design must be environmentally holistic.

A holistic approach takes into account the entirety of the systems and functions of the ambient environment. The built system itself has inorganic and organic components and will have various impacts on the systems of the environment; however, ecological design must not isolate any one environmental function or element, but rather see the ecosystems as a functional whole. It is the responsibility of the designer to make certain that the existing ecosystem survives the introduction of the foreign mass of the built system intact, and that no part of the ecosystem is irrevocably damaged or destroyed, unless all contingencies have been taken into account and corresponding preventive measures have been instituted.

● Ecological design must be an anticipatory design approach.

Ecological design endeavours must be responsive and anticipatory strategies because environmental impact is inevitable. By its presence, the built environment will cause changes, such as depletion, to the ecosystem; its construction and functioning will necessitate the consumption and redistribution of natural resources.
 However, alteration of the ecosystem through human activity is not necessarily destructive or even undesirable. It would be a mis-

understanding to think that 'green' design means no design – that the whole biosphere should be off limits to any kind of development. Ecosystems evolve over time even in the absence of human intervention, so it would be impossible to set as a goal the prevention of any kind of environmental change. Ecological design does not therefore aim to preserve the biosphere from human influence, but to relate human intervention in ecosystems to the environment in a manner that does the least damage, which, as we have seen, means recognizing the limits of the ecosystems themselves. Only in this way can we make our intervention by the designed structure as beneficial to the ecosystems as possible, and with proper regard to the anticipation of changes which will necessarily be wrought on the environment by the presence of the building.

● Ecological design is multi-disciplinary.

The designer's ecological solutions are usually multi-disciplinary. Environmental problems arising from human activities in the natural environment arise from stresses caused by the intervention, and can take one or all of the following three forms: depletion, alteration, or addition to the earth's ecosystems and resources (both globally and locally). Ecological design methods try to minimise the adverse impact of human interventions (built structures) and to reduce as much as possible harm to the ecosystem. It is up to the designer to, from the outset, anticipate adverse results that flow from the creation of the building and minimize them in the design process, as well as making it a priority to ensure that the elimination of negative impacts on ecosystems and natural resources continues – in essence, that the ecological approach is institutionalized.

An inter-disciplinary, ecological approach to building design is crucial. It must encompass, obviously, ecology and architecture, but also other related disciplines, such as engineering, chemistry, and materials sciences, that are also concerned with the problems of the protection, conservation and preservation of the environment. Unfortunately, most of today's designers lack the knowledge of ecology and biology required by 'green' design. A further problem is that an agreed definition of what constitutes ecological architecture has yet to be produced, along with a 'green' design theory embodying a set of generally recognized principles. Because ecological design is as yet undertheorized, the present work seeks to provide a sound and comprehensive theoretical framework and unifying principles, in the absence of which ecological design will remain partial, misunderstood, and inconsistent (see chapter 3).

The designer's current range of expertise needs to be critically expanded. The principles of architecture have to be widened to include an ecologically sensitive analysis of the building's construction, operation during its useful life, and final disposal.

The need for ecological design has been given attention in recent years (e.g. Moorcroft, 1972; Commoner, 1972; Marras, 1999, et al.), but an awareness of the nexus of architecture-ecology, reflecting the intimate relationship of structure and environment, must become an integral part of the practice of architecture – beginning with its incorporation into architectural pedagogy.

More than a decorative art, architecture is a social practice which involves various disciplines, each of which makes a particular contribution to it. Ecology, a science which deals with the living systems of the biosphere and their many components – of which human beings are one – needs to be included among the disciplines that contribute to architecture. Human beings are constantly modifying the environment, and their activities are one focus of ecology. The creation of a built environment within the natural one is a paramount way in which human beings affect ecology, often impairing the natural systems already in place by the imposition of manmade ones (the impact of urbanization is a chief example). It would go a long way toward minimising the ecological impact of architecture and development if architects and designers were sufficiently trained in ecology, so that conflicts between architecture and nature, between the built structure and the ecology of its environment, could be mimimised and negative impacts reduced.

Architecture today, however, takes an 'either/or' approach rather than the holistic one that is being described here. The 'spatial' (e.g. Martin, 1967) and the 'climatic' approaches tend to dominate architectural theory and practice at the present time (in Hillier, 1977). The spatialist school stresses the degree to which any constructed object (the building) takes up space in the environment and thereby modifies it. The climatic approach recognizes that the building, as a system which 'breathes' or interacts with the surrounding environment, changes that environment and particularly its climate – being a key defining limit on the ecosystem (as well as on the internal climate of the built structure). 'Green' design, however, includes both aspects. The built environment as a 'system' and the earth with its ecosystems as the 'environment' must be considered simultaneously in the ecological approach to design (see chapter 3).

The climatic approach begins to take acount of the element of 'feedback', the manner in which the outputs of the built structure into the environment cause ecological changes and therefore result in new inputs to the built structure. But the range of ecosystemic interactions between the building and its environment is far broader than that taken into account by current architectural practice and pedagogy. Beyond the purely physical spatial displacement caused by the insertion of the building into an ecosystem, there are systemic interactions between building and ecosystem which the designer must take into account and incorporate into his design. These interactions begin with the choice of materials to be used in the structure, and the energy source to be used to make the built system run. The functioning of the building depends on a set of internal processes, the building's 'metabolism', which interact with the environment (into which, of course, their wastes and exhausts are discharged). The architect has to be aware of these processes and their effects, as well as the ecosystem's response to them.

It is clear that the continued unabated conversion, simplification, and degradation of the earth's ecosystems need to be reversed. Degraded habitats should be restored so that they can perform critical functions. Examples of steps needed in this direction include using artificial wetlands for flood control and nitrogen abatement, and promoting reforestation for watershed protection and carbon sequestration.

We can no longer assume that nature's beneficial 'services' will always be there free for the taking. We must become more cautious and forward-thinking before taking any actions that disrupt natural systems and limit the options of future generations. We have already seen that the degradation of the ecosystem can have severe economic, social, and environmental costs, even though we can only measure a fraction of them at present. We can rarely determine the full impact of our actions; the consequences for nature are often unforeseen and unpredictable. The loss of individual species and habitat and the degradation and simplification of ecosystems can impair nature's ability to provide the functions on which our lives depend. Many of these losses are irreversible, and much of what is lost is simply irreplaceable. Ecological design has to be based on the fact that nature's processes, regenerative capability and resources are limited and that conservation is essential.

Maintaining nature's systemic viability – as the good health of the processes that are so important to us – requires looking beyond the needs of this generation, with the goal of ensuring sustainability for many generations to come. Thus, ecological design has to be anticipatory. We must act under the assumption that future generations will need at least the same level of nature's services as we have today – the safe minimum standard. Thus reason and equity dictate that we operate under the precautionary principle. We can neither practically nor ethically decide that future generations can simply do without.

What can be done to stop the unraveling of nature's life-support system and ensure that it can continue for generations? First, our understanding of the true extent and value of nature's functions, and the tools and processes we use to make decisions, need to be redirected toward ensuring the sustainability of the planet's life-support system. Understanding and valuing nature and ensuring that its resources are used equitably and within the finite limits of its regenerative capacity is essential to sustainability.

To summarise, we need to appreciate the interconnected web of life that we are part of and that supports us, both locally and globally within the natural systems of the biosphere. We must realise that the cumulative impact of our activities on one location can have an impact elsewhere on another location. Ecological design must embody ways for man to use ecosystems that capitalise on nature while at the same time maintaining its stability, resilience and productivity. For example, by maintaining a nearby forest and estuary, a diversity of crops, and a variety within each crop, farmers are assured a sufficient harvest regardless of weather and pests. This may not yield the maximum under 'ideal' conditions (which rarely exist), but it is smart 'crop insurance'. Similarly, many human societies have evolved strategies for not only coping with nature's inevitable rhythms and changes, but also for using those changes and 'disturbances' to their benefit, such as flood-dependent agriculture and flood-plain fisheries. The bottom line is that for humans to be healthy and resilient, the natural systems must be so, too.

This chapter has dealt with some of the key aspects of the environmental properties and processes that are crucial to understanding ecological design. We can summarise the main assumptions that underlie our ecological approach to design as follows:
● The basis of a sustainable future is the knowledge that it is in the interest of humanity to maintain local and global ecology in functioning and viable condition. This implies limiting as far as possible the destructive effects of human systems and designs on ecosystems.

● The current pace at which human beings are destroying global ecosystems is non-viable – which is why human actions (including architectural design) have to become ecosystem-sensitive.

● Natural resources are limited. Waste, once it is produced, is not easily recycled. Design must be regarded as conservation of resources.

● People are part of a closed system in the biosphere, and the processes of the natural environment, being unitary, must be considered holistically as part of the design and planning process in the creation of the built environment.

● There are interrelationships and interconnections between the manmade environment and the natural environment both locally and globally. Hence, any changes to any part of any one of these systems affect the entire system. Design must be regarded in terms of connectivity of global and local ecological processes and resources.

The design objectives discussed here are fundamental and vital to our ecological approach; these basic premises need to be acknowledged in all design assignments and constructive undertakings. Furthermore, there is a need to establish for the designer a more general theoretical framework for ecological design which will unify all these elements into a holistic design model (see chapter 3).

The key points to note here are that ecological design is a complex endeavour, and that design should in effect be a form of applied ecology, where a proper and thorough understanding of the ecosystem of the project site for our proposed building and its relationships with the biospheric functions and global resources are essential. To be environmentally holistic, the designer has to regard his built system as a set of connected interrelationships and interactions with the natural systems in the environment.

We can define ecological design as the prudent management of the holistic connections of energy and materials used in the built system with the ecosystems and natural resources in the biosphere, in tandem with a concerted effort to reduce the detrimental impact of this management, thus achieving a positive integration of built and natural environments. In addition, we need to ensure that this endeavour is not a once-only effort; the interactions of building and nature have to be monitored and managed dynamically over time (i.e. in the entire life cycle of the built system from source-to-sink and encompassing the totality of its use of energy and materials).

Chapter Three

A Theory of Ecological Design

For ecological design to be durable, we must have a theory – a general theoretical basis that will enable our design work to be environmentally holistic and anticipatory. What is generally found to be inadequate in current theoretical constructs is their incompleteness or failure to include an environmentally holistic property (e.g. "connectedness") crucial to the ecological approach.

This chapter will establish the fundamental criteria for ecological design and show how these are all interrelated. Therefore, any approach to design that does not take into account these aspects or the interactions between them cannot be considered holistic, and hence must either be not ecological or at best an incomplete approach to ecological design.

To begin with, we should first acknowledge that ecological design is complex, and certainly considerably more complex than is currently recognised by many ecological designers today. More specifically, it involves the incorporation of a complex set of "interdependent interactions" or connections with the environment (both global and local) which furthermore must be regarded dynamically (i.e. over time). This explains the need for establishing the holistic and anticipatory properties of ecological design.

To achieve environmentally sustainable objectives, ecological architecture must minimise (and at the same time be responsive to) the negative impacts that it has on the earth's ecosystems and resources. As mentioned earlier, we should be aware that ecological design is not a retreating battle, but on the contrary a designed system that can contribute productively to the environment (e.g. through production of energy using photovoltaics) as well as restore and repair damaged ecosystems.

A General Systems Framework for Design

For the purpose of developing a theory for ecological design, we can regard our building as a system (i.e. a designed system or a built system) that exists in an environment (including both the manmade and natural environments) (see Fig. 17). The general systems concept is fundamental to the ecosystem concept in ecology. Briefly stated, in the analysis of the relationship of any system to its environment, there is essentially no limit to the number of variables that we can include in the analysis or in the description of the design problem. In fact, this applies to all design endeavours, for no matter how fortunate our choice may be of inputs and outputs to describe a system, they cannot be expected to constitute a complete description. The crucial task in design – and similarly in any theory – is therefore to pick the right variables to be included, which are those we find essential to our resolution of the design process (see Fig. 18).

What we need is a simple, general framework that structures the entire set of ecological interactions between a designed system and the earth's ecosystems and resources. This framework must be able to identify for the designer those impacts that are undesirable and which need to be minimised or altered in the design process. The theoretical basis for ecological design must provide the designer with a set of structuring and organising principles to carry out this goal. It can take the form of an open structure with which the selected and relevant design constraints (e.g., ecological considerations) can be holistically and simultaneously organised and identified. Furthermore, the open structure must facilitate the selection, consideration, and incorporation of design objectives in our subsequent design synthesis.

The open structure can be simply a conceptual or theoretical framework and should enable the designer to decide which ecological considerations to incorporate into his design synthesis, while at the same time ensuring a basis for a comprehensive check of other interdependent factors influencing design. Crucially, it must also demonstrate the interrelationships of all factors, which is an essen-

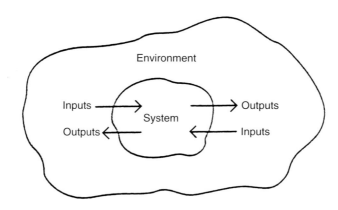

Fig. 17 Model of a system and its environment and the exchanges between the two (Source: Yeang, 1995)

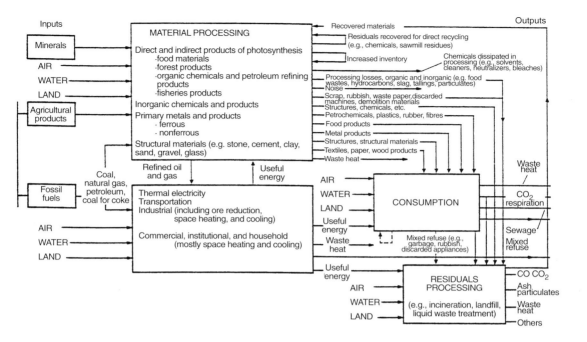

Fig. 18 The inputs and outputs through the built environment (Source: Kneese et al., 1971)

tial property of the connectedness of all ecological systems in the biosphere. By using the open structure as a design map, the designer can also include any other related and pertinent disciplines that are similarly concerned with the problems of environmental protection and conservation in design solutions (e.g. waste disposal, resource conservation, pollution control, applied ecology, etc.). Thus, the essential properties of the theory of ecological design are that it be inclusive, comprehensive and open.

Green Design as Environmental Impact Statement

As mentioned earlier, ecological design is prognostic and anticipatory, and as a result the design process becomes essentially the preparation of a "statement" of anticipated environmental impacts and benefits. From the earlier examination of ecology and ecological concepts, we have determined that the extent of the environmental consequences of any built system can be seen as the net result of its dependencies (i.e. demands and contributions) on the earth's ecosystems and processes and on the earth's energy and material resources (e.g., for a specific product or a specific function; see chapter 2). These dependencies are both global (for example, the use of non-renewable resources) as well local (the impact on

local ecology). If the designer is aware of the ecological consequences (both detrimental and beneficial) of his design, then this
knowledge represents in effect a summation of the extent of the
design's impacts on the environment, which thereby are accepted
and anticipated by the designer.

Defining the design task in this way should not imply the exploitation of the biosphere (Bookchin, 1973). On the contrary, this
approach emphasises the extent of human dependency (and of
built structures) on the earth's resources and ecosystems. Such a
viewpoint helps direct our attention to those aspects of the
designed system that have ecological implications and indicates
the critical areas where the undesirable impacts might be eliminated, reduced or remedied. The ecological approach embodies the
realisation that any designed system is dependent directly or indirectly upon the biosphere for specific elements and processes,
which can be identified (as in chapter 2) as including the following:
● Renewable and non-renewable resources including minerals, fossil fuels, air, water and food.
● Biological, physical, and chemical processes, for example decomposition, photosynthesis and mineral cycling.
● End point or processing of waste and discharges resulting from
human activities, including life processes as well as the functioning
of manmade systems (example: landfill waste disposal).
● Physical space in which to live, work and build.

These various functions and aspects of the environment and our
human use of them are interrelated and overlapping; they meet
each other at 'transfer points', where the designed system interacts
with the surrounding ecosystem. Transfer points are vitally important to green design precisely because bad design at the points
where exchange occurs frequently results in damage to the ecosystem.

It should be remembered, however, that it is impossible to
design a system in which none of these linkages results in an
impact on the ecosystem. The mere physical existence of the building, as we have seen, causes some spatial displacement (it takes up
space) of the ecosystem, and our use of land represents a loss to the
biosphere volumetrically. Absolute ecological compatibility is physically impossible because of this most basic impact on the environment. However, we can produce buildings that have greater or
lesser destructive effects on the environment, and they can even
have some results that are beneficial. It is the job of green design
to minimise negative impacts and maximise beneficial interactions
between built systems and natural ecosystems.

Dynamic Aspects of the Theory

From the ecologist's point of view, architectural design results in a built form (with attendant operational systems) that can be equated to a net statement of its physical and potential demands and influences on the ecosystems and on the earth's resources. A complete statement of this kind necessitates tracking the flow of natural resources (including energy) into the designed system from source to sink, which is to say from their extraction from the environment through their employment in the built system and back into the environment as output or waste (viable now by using digital technologies). If we accept this principle, then we must consider every aspect of the built environment, including its physical, economic, cultural and other roles, in terms of the structure's relationship to the environment. Furthermore, this relationship has to be viewed over the entire course of the structure's useful life. The life-cycle concept is crucial to ecological design. As designers, we must try to identify the ramifications of every design decision, identifying undesirable ecological impacts and taking appropriate action to minimise or prevent them. In practice, it is not possible to identify nor to quantify these in their totality, but indices can be used and a broad theoretical framework can be developed to reveal the interdependencies.

The Theoretical Structure of Ecological Design

The theoretical framework of green design must be developed in line with a number of concepts, which will be briefly summarised here.

A building exists both in terms of its physical being (form, siting and structure) and its functional aspects, i.e. the systems and operations that sustain it during its useful life. Both aspects involve the built structure in relationships with the natural environment which take place over time. The building acts like a living organism; in place of food, it uses of energy and materials, and also produces outputs into its environment. Our theoretical structure should model these exchanges.

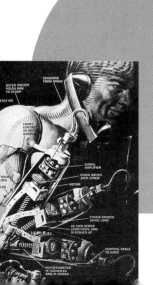

Three components are essential for an ecological model of the designed system. Our framework must include a description of the built system itself, a description of its environment including the ambient ecosystem and natural resources, and a mapping of the interactions between these two components (i.e. between the building and its environment).

The first step is to systematically take account of the internal processes of the designed system. The second step is to measure, based on a thorough knowledge of the building's physical and functional requirements, its interactions with the earth's ecosystems in the form of the energy and resources removed from the environment by the construction and ongoing operation of the structure, and in addition amounts of matter and energy that are sent back into the natural environment as a result of the functioning of the building's internal systems (the 'metabolism' that makes it function as a built environment).

A supplementary question is the relationship of the built structure as an element in the spatial configuration of the environment. Its existence as a built environment within the natural one implies further interactions and effects on biosphere. Analysis of any such impacts will also have to be factored into the theoretical framework.

An open general systems framework
can be used to visualise 'sets of interactions'
taking place between the designed system
and its environment.

The concept of an open system in contact with its environment as formulated in general systems theory is useful here. Based on the above analysis of the fundamental interactions of the built and natural environments, the interactions can be grouped into four sets.

Set 1: External interdependencies, consisting of the designed system's relations to the external environment.

Set 2: Internal interdependencies, being the designed system's internal relations.

Set 3: External-to-internal exchanges of energy and matter – i.e., system inputs.

Set 4: Internal-to-external exchanges of energy and matter – i.e., system outputs.

The four sets also usefully describe the 'transfer points' between the built and natural environments which were discussed above. Green design must take account of all four sets as well as the interactions between them. In this way, our framework allows us to determine how architecture impinges on terrestrial ecosystems and natural resources whenever we address any design task.

Elsewhere (Yeang, 1995) I have developed a 'partitioned matrix' (LP) which unifies these sets of interactions in a single symbolic form. The figure demonstrates this conceptualisation of the relationship of the designed system and its environment (suffix 1 stands for the system, suffix 2 the environment). If the letter L stands for the interdependencies within the framework, then four types of interaction can be identified (Tolman and Brunswick, 1935; Emery and Trist, 1965; Walmsley, 1972). In the partitioned matrix, they are identified as L_{11}, L_{12}, L_{21}, and L_{22}:

$$(LP) \quad = \quad \begin{array}{c|c} L_{11} & L_{12} \\ \hline L_{21} & L_{22} \end{array}$$

Remembering that '1' represents the built system and '2' the environment in which it is situated, we can map the four sets of interactions listed above onto the partitioned matrix. L_{11} represents processes that occur within the system (internal interdependencies), L_{22} represents activities in the environment (external interdependencies), and L_{12} and L_{21} refer to system/environment and environment/system exchanges, respectively. Thus, internal and external relations and transactional interdependencies are all accounted for.

The partitioned matrix is itself a complete theoretical framework embodying all ecological design considerations. The designer can use this tool to examine interactions between the system to be built and its environment holistically and inclusively, taking account of all the environmental interdependencies described by the above four sets.

The Law of Ecological Design

If a fundamental "law" for ecological design can be asserted, then this partitioned matrix constitutes the Law of Ecological Design. In ecological design, this "law" then requires the designer to look at his designed system in terms of its component parts, i.e. inputs, outputs, and internal and external relations, and then to see how these interact with each other (both statically and dynamically over time, these being the four components of the partitioned matrix).

In effect, the designer can then further ascertain which of their ecological impacts need to be given priority and which need to be taken into account or adjusted in the process of design. In this way, any designed system can be conceptually broken down and analysed based on these four sets of interactions as follows:

L22

These interactions describe the designed system's external interdependencies or 'external relations'. By this is meant the totality of the ecological processes of the ambient ecosystem, which as we have seen interacts with other ecosystems; hence L22 takes in not only local but global environments and terrestrial resources in their totality. It therefore also includes the processes by which earth's resources are created (for example, the formation of fossil fuels and non-renewable resources), which may be affected by, and themselves affect, the built structure's functioning. These external resources will be modified, depleted or added to by the creation and functioning of the built system.

L11

The internal interdependencies are the internal environmental relations of the built system. This means the sum of all the activities that go on inside the building, including all of its operations and functions. The functioning of the built structure's internal metabolism will have larger effects, extending to the ecosystem where it is sited; these effects, by the principle of connectivity (see last chapter), will in turn affect other ecosystems and the biosphere's totality of resources. The L11 effects take place over and describe the whole life cycle of the building.

L21

This quadrant of the matrix describes the total inputs into the built system, including all of the exchanges of energy and matter that go into its construction. System inputs of a designed system include all of the resources that make up its component parts and the matter and energy upon which its operations and processes depend. Securing these resources that make the building 'run' (the extraction of infrastructural materials and energy from the earth) often causes damage to the biosphere and its ecosystems.

L12
The total outputs from the built environment into the natural one
are the most obvious concern of the ecological designer, but they
are only one quarter of the total interactions discussed here. These
outputs, however, include not only discharges of waste and exhaust
from the building's construction and operation, but also the physical matter of the structure itself, which must be disposed of at the
end of the building's planned lifespan. Obviously, if these outputs
cannot be assimilated by the natural environment they result in
ecological harm.

Any design approach that claims to be ecological and does not take
into account these four components and their interactions over
time cannot be considered a complete and holistic ecological
design, as interconnectedness is a crucial characteristic of ecosystems. Failure to take this factor into consideration is non-eco-
logical.

Design Implications

As stated earlier, holistic and ecological design takes into account
local and global environmental interactions; anticipatory design is
forward-looking and is also environmental in that it considers
effects over the entire lifetime of the built structure. A further
point is that green design is self-critical. It considers its own effects
on the environment and tries to eliminate negative impacts on ecosystems and terrestrial resources. An ecological designer works
within the constraints set by principles shown here, and tries to
maximise the utility and efficiency of the design while reducing
negative effects of the building's creation and functioning. Thus,
the green designer takes a 'balanced budget' approach, weighing
environmental costs and using global resources in the least damaging, most advantageous manner possible.
    The ecological design framework provides a structure for identi-
fying linkages among environmental elements. Energy consumption, waste generation, resource use in the production of building
materials, transportation of building occupants, operation of building services and systems, and other processes in the life cycle of a
building can be linked with one or more changes in the quantity or
quality of environmental components. The cascading effects of
these changes can then be traced through to their effect on ecosystems and specific communities – the end points of environmental
concern. We are then able to evaluate the relative importance of

various potential effects on the productivity and viability of plant and animal communities.

When we look at a specific environmental effect, we look at the entire chain with which it is associated and we avoid focussing only on intermediate impacts. The linkages among the effects are not simple linear chains but complex webs; each emission or use of a natural resource can cause a number of changes in the quality or quantity of air, water, soil and resource stock. In turn, these changes in environmental quality and resource quantity can affect different end points.

From the point of view of applied ecology, ecological design has essentially to do with energy and materials management concentrated in a particular locality (i.e., the building site). By this I mean that the earth's energy and material resources (biotic and abiotic components) are in effect managed and assembled by the designer into a temporary manmade form (for a period of intended use or 'its useful life'), then later demolished or disassembled at the end of this period, to be either reused or recycled (see chapter 4) within the built environment or assimilated elsewhere into the natural environment. But, however mechanistic this formula may seem, we must be clear that ecological design is more than simply the management of energy and materials. The designed system must create a balanced ecosystem of biotic and abiotic components or, what would be better, create a productive and even reparative (i.e., healing) relationship with the natural environment both locally and globally. Of course, in addition one has to consider the other conventional aspects of the design of a built system (in this case, the skyscraper): design programme, costs, aesthetics, site, and so forth.

The green theoretical framework reminds the designer that his or her building is not just a spatial object, but that its internal functions (L_{11}) and external relations (L_{22}) are equally part of the designed system. Environmental interactions must always be taken into consideration by the design process, and not only the physical reality of the structure and its components; the functional aspects of the building and its outputs over its useful life and the subsequent disposal of the structure itself are also parts of the partitioned matrix.

The partitioned matrix and the above breakdown of sets of interactions can be used by the designer to conceptualise, verify and check the environmental interactions and effects of the structure in question. But the architect can also use the framework to analyse any design or 'deconstruct' it, separating out its component interactions into the quadripartite system of the matrix: resource inputs (L_{21}) and outputs (L_{12}), which are dealt with further in chapter 4; internal functions (L11), treated in chapter 6; and external environmental relations (L22), which are the subject of chapter 5. It is worth underscoring that the architect's ethical responsibility extends over all the aspects of the matrix (i.e., it includes the totality of interrelations), which must be factored into the design. Beyond these design imperatives in the narrow sense is an overarching responsibility to keep 'green' principles always in mind, including the general aim of sustainable development (the ecosystems must remain viable after our intervention) and minimisation of the damaging side effects of human action on the environment. As has been said elsewhere, human interventions can even be made mutually beneficial for the built and natural environments.

A further aspect of the matrix is that, because it is systemic and comprehensive, it provides a check on environmental impact assessments. It reminds the designer of the scope of the anticipated impacts and interactions of his design. For example, it could happen that an architect could overlook one interaction, or emphasise the importance of a single factor at the expense of another, thereby producing lopsidedness in the design.

If a designer is particularly concerned with pollution (negative outputs) from the built system, even though this is a 'green' objective, the drive to reduce these outputs could result in a design that required excessive energy inputs, thereby consuming more terrestrial resources and perhaps putting strain on other (possibly non-local) ecosystems. The use of the partitioned matrix prevents this 'seesaw' effect by requiring the designer to keep in mind that any design that does not deal with the totality of environmental interactions and ramifications over the building's entire functional life will be uncomprehensive and therefore environmentally unsatisfactory.

The matrix framework will also require the architect to integrate the building's systems and components into the ambient ecosystem in such a way as to achieve compatibility, and even symbiosis, with the structure's site. But one thing the matrix will not do is to incorporate the environmental feedback that occurs once the building is actually constructed. Once the designed system is physically in place, its outputs will have a greater or lesser environmental impact, which could reduce the ecosystem's ability to provide the inputs and natural resources specified in the design (Miller, 1966). A more comprehensive and complex model that would encompass the feedback loop must be further developed from the framework as it now stands.

The key feature of this theory is its comprehensiveness. Previous definitions of ecological design (e.g., Wells, 1972, etc.) have been applied, without much precision, to any design method that voiced some environmental concern, but without providing any check on their validity or comprehensiveness. Thus, this theoretical framework can be used as an analytical tool to evaluate the 'eco-logical' design approaches of other designers and to test their comprehensiveness. Since the ecosystems approach as it is here put forth is comprehensive and synoptic, those approaches that in one or more ways give a truncated version of the framework, leave out quadrants of the matrix or otherwise fail to notice certain environmental interdependencies will not qualify as truly holistic, forward-looking and 'green'. As I have shown above, an incomplete environmental approach is as capable of producing ecosystem damage as a pre-ecological one, and may actually add to the very sort of problems it aimed to avoid. The validity of the matrix and the framework outlined here rests, therefore, on its comprehensiveness.

To summarise, the interactions framework described in this chapter has four prime functions:

● The architect or designer is given a conceptual framework for organising and coming to grips with all the ecological ramifications of the proposed structure, in this case the skyscraper and other similar intensive building types. After identifying the totality of interactions between structure and ecology, the designer is able to minimise negative environmental impacts through design decisions relating to various factors such as materials to be used and their assembly into the built structure.
● The model can be shared, i.e. it can be used as a common frame of reference for the designer and other professionals in other disciplines who are evaluating the ecological impact of the designed system (the skyscraper). This commonality promotes a 'multiple comprehensiveness' in that the examination of interrelated envir-

onmental issues is followed through in a continuous, harmonious manner.

● Over time, the establishment of a common frame of reference through this model creates the possibility of further theoretical elaboration. For environmental concerns to be resolved, various fields with similar concerns, heretofore separate, must be brought together. For example, efforts to conserve natural resources or provide alternatives can be contributory to the design process.

The real test of environmental commitment and principles is on the level of human action (i.e., when ground is broken), and this model, by offering a comprehensive framework for understanding the interrelations of built systems and ecosystems, allows people in various fields to act in concert and contribute to ecological design philosophy.

The theory of interactions presented here provides a single, unifying theory to bring together under one umbrella aspects of environmental science and protection efforts that have in the past been uncoordinated.

● Just as it can unite disciplines, this design model can be extended to other disciplines. The designer can use the framework to describe and anticipate the environmental impact of a skyscraper, but other theorists and practitioners might employ it to model a broad range of human activities with ecological consequences, for example tourism and the effects of recreational uses of natural sites.

● Lastly, the theoretical structure developed here points out the holes in current design practice and research on the subject. Green design, when pursued comprehensively, demands certain kinds of data, which will have to be developed and quantified where not available. This comprehensive design framework offers a benchmark, which can be used by architects and designers to evaluate on a consistent and quantitative basis any proposed design, or to compare one design with another.

The above premises provide the broad theoretical basis for ecological design, applicable to our skyscraper and other intensive building types. From this starting point, a strategy for practical application of ecological design of the large building type should begin by first addressing its design in terms of energy and materials conservation (i.e. L21 and L11), or more precisely in terms of the management of energy and materials in the designed system throughout its entire life cycle (taken up in chapter 4). This is the obvious beginning for the design process, since current figures show that over 28 percent of national energy use is in buildings, and that over 50 percent of a nation's wastes that go into landfills comes from buildings. Significant ecological benefits can be achieved by the low-energy design of intensive building types, and through concerted design efforts that seek to achieve cyclic use of materials within the built environment. We need to ask questions about the materials that are used in all components of the skyscraper (e.g. its piling, structure, facade, infill walls, finishing materials, operational systems, etc.).

We must be clear that the interactions framework is not a substitute for design invention. Invariably in any designed programme, the designer has still to synthesise the selected set of considerations into a physical form. Our building has still to be designed, though obviously based on informed decisions. In this process of synthesis, the structural model described here is useful in determining ecological interactions and implications. Such design decisions are of course in addition to the usual architectural and engineering decisions that need to be made in the design of the skyscraper and other large buildings.

The partitioned matrix also points out the fact that design decisions and materials selection can have impacts on ecosystems away from the project site. Every design problem represents a particular ecological balance in the dimensions of the relative importance of its principal elements and the demands arising therefrom; in each case, the synthesis of a design that is related to that balance would appear to be the most effective way of designing an ecologically responsive built environment.

Different design methods might be viewed as alternatives which are more or less advantageous depending on the design problem at hand (and on the designer as well). Our intention should not be to try to predetermine a set of standard solutions for design, for no single solution or set of solutions could be sketched out that would automatically correct all environmental problems. The aim here is not to provide a panacea (an impossible task, given the diversity of designs and situations), but rather examples and options to provide the designer with insights into aspects of environmental impairment by built systems. In some instances, solutions may arise which do not require the synthesis of a physical system at all. The technical solutions discussed here are simply what is currently the state of the art (see chapters 4 and 6).

The eventual impact of a building will reflect the degree to which the architect was able to take in the whole spectrum of environmental effects during the design process. Yet, while the partitioned matrix is a comprehensive framework, it is not programmatic. That is to say, it includes all possible issues but not, for obvious reasons, particular situations and cases. It can act as the 'law for ecological design', but it is the individual designer who has to apply that law. All that can be predicted here is the type of design issue likely to be faced by the architect of a 'green' skyscraper and other large buildings, particularly in the area of ecosystem interactions and effects. The framework of green design has to be instituted from the outset because initial choices will largely determine the degree of environmental damage, the magnitude of the feedback effect into the designed system and the possibility of correction.

Even before the design begins to take shape, the designer must have analysed his or her strategic options under the partitioned matrix and the green design framework. The results of this analysis will have the effect of narrowing down the range of solutions and making clear the interrelationships which need to be understood in solving the design problem. The various factors and relationships might be represented graphically so as to highlight their most important features; this might also result in a schematic that would be general enough to be used in a multiplicity of cases, i.e. real design tasks and projects and not just idealisations. Thus, similarities or differences in design issues would be highlighted.

For now, the interactions model and the matrix present a general, overall picture of the design problems faced by architects following green principles. In essence, it is a map, which allows many paths on the way from problem recognition to resolution. How the architect goes about negotiating the obstacles and constraints depicted on this map of the design task is personal, both in terms of the designer's individuality as an architectural practitioner and also the specific environmental and other factors of the site to be used. What is important is that in adapting the built system to the natural environment the designer does not neglect any of the interactions defined by the partitioned matrix; how they are addressed remains individual.

Having developed the theoretical foundation presented in this chapter, it is now possible to look at particular solutions to design

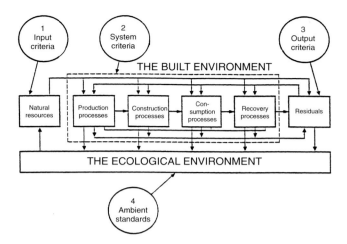

Fig. 19   The built environment (Source: Yeang)

1. Input Criteria
   - Quantities of energy and materials used in the designed system
   - The availability of the energy and material resources (rates of depletion)
   - The ecosystem consequences of each input used

2. Output Criteria
   - The permissible quantities of output discharged by the designed system
   - The routes taken by various outputs after discharge and their ecosystem consequences
   - The energy and material cost of management of the output
   - The ecosystem consequences of output management

3. System Criteria
   - The extent of the pattern of needs and use
   - The efficiency of the system processes
   - The extent of internalization of the system processes
   - The ecosystem consequences of the realization of the designed system

4. Input Criteria
   - Carrying capacity or extent of resilience of the natural systems in the environment

problems on a technical level. These also mirror the partitioned matrix and the interactions depicted in figure 19:
- the management of inputs, or L21 (see chapter 4),
- the management of outputs, or L12 (chapter 4),
- the management of the environmental context of the building, or L22 (see chapter 5),
- the design and management of the internal operational systems of the skyscraper in relation to the other three sets of factors, or L11 (see chapter 6).
- the interactions of all the above sets acting symbiotically as a whole with the natural systems (and other manmade systems as well) in the biosphere.

To fulfil the last (and broadest) goal, that of synchronising all the above aspects of the skyscraper (i.e. its inputs, outputs, operational activities and environmental consequences) with the natural cycles in the biosphere and with the other human structures, communities and activities in the biosphere, appears at first to be naively idealistic. Yet it is crucial to the realisation of green design and sustainability; however, it would require economic-political decisions that lie beyond the scope of the present work and full exploitation of the opportunities in the use of digital technologies.

Chapter Four

Assessing What Is to Be Built

Re-evaluation of the User's Needs

Ecological design is like conventional building planning in that one must begin by asking what is to be built and if it can be built. With any commercial building, at the onset of the process we have to ask whether the project is financially viable before the designer begins designing a structure; in the same way, the ecological designer must at the very start assess the set of design requirements and whether the requirements are also ecologically viable.

Essentially, the general spatial and systemic requirements of the skyscraper that need to be re-evaluated prior to design are as follows:

- Spatial (gross floor area or GFA)
- Building footprint
- Building configuration (e.g. high-rise, low-rise, mid-rise, etc.)
- Internal environment conditions
- Operational consumption level (inputs, outputs, etc.)
- Transportation implications

Right from the start, the designer must place greater priority on the evaluation of the project's users (people) and their needs than on responding to the demands for provision of hardware (e.g. built enclosures, equipment, mechanical systems, etc.) (see Fig. 20). The tendency of most designers is to do the opposite. While the scale in the design of the skyscraper may appear daunting, we will find that in many respects, this initial strategy for design will apply to all projects whether big or small.

Rather than taking the client's demands for granted, we need to look into ways in which the designer and others might inadvertently contribute to creating what may prove to be unsustainable expectations regarding the nature and form of our built environment. If the built system is a required commodity, then the first step is to assess the extent of building to be done and how to make this ecologically responsive. The objective of ecological design is to ensure that the building can be designed in such a way that a minimum of non-renewable sources of energy is needed to service the structure in terms of heating, hot water, cooling, lighting, power, ventilation, and other internal functions.

The Spatial Requirements

In the design of the green sky-scraper and other intensive building types, as with any other design problem, the designer has first of all to evaluate the extent of design requirements for the building. Essentially, this means an evaluation of the design brief in terms of its level of enclosure (e.g., the extent of gross areas to be enclosed) and the level of environmental systems to be provided, as well as other factors, all of which are related to the skyscraper's consumption levels (i.e. L21 in the partitioned matrix) and emission levels (L12).

While there are commercial justifications for achieving maximum plot-to-GFA (gross floor area) ratios, as in most high-rise conditions, the extent of enclosure may vary. The designer needs to assess the extent of total enclosure, the provision of partial enclosure or the provision of transitional areas (e.g. skycourts) within the built-form.

- Human welfare
  is related to
- Material standard of living
  which depends on the
- Provision of manufactured goods
  which in turn requires
- Consumption of natural capital
  which means
- Extraction of natural resources
  which involves
- Discharges of waste
  which result in
- Pollution

Ecological design essentially requires the establishing of standards and determining the extent of user requirements that is acceptable to society. For instance, if we are to reduce the volume of our discharges of waste energy and materials to the natural environment, we need to reduce the extent and rate of our depletion of natural resources. To do this means that we have to reduce the consumption of natural capital and the provision of manufactured goods. To reduce the provision of manufactured goods would lead to a reduction in the material standard of living enjoyed by the developed countries. Lowering the standard of living will lead to lowering welfare for the human communities. The question then becomes a subjective one: what are we prepared to sacrifice to have a sustainable future? The designer's role is to use innovation to help achieve one.

Fig. 20   Ecological design (Source: adapted, Reid)

It is contended here that most project sites for skyscrapers are essentially zeroculture ecosystems (further discussed in chapter 5), in which case we need to rehabilitate the site with as much organic mass as possible. Ideally, the equivalent built-up space that is designated as GFA should be given to terraced planted areas, preferably at a greater than 1:1 ratio.

The Building Footprint and Building Configuration

The determination of the floor-plate size and the built form determines the footprint on the site and hence its impact on the physical features, such as vegetation and hydrology/groundwater. The structure's form will create shadows on surrounding land that will change over the course of the day and the seasons of the year, and will also influence the wind conditions at all levels of the surrounding streetscape and buildings. As a general strategy, we should build on a minimum site area using small footprint designs and leaving large portions of the site undisturbed.

Of course, the ecological quality of the site itself must be taken into consideration (see chapter 5). We must avoid placement of buildings in those areas within our site where development is likely to have a significant environmental consequence.

The Internal Environment and Operational Consumption Levels

Deciding on the internal environment and the operational consumption levels of the skyscraper is probably the most important decision affecting energy use. In the conventional building, these elements take up to 65 percent of the energy used in a building over its life cycle. These levels are often locality-specific and related to the standard of living and welfare of the population of the project site. The extent of shelter and comfort required by the people

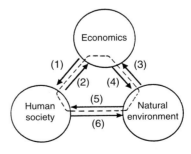

Fig. 21   Conceptual diagram of man-environment relationship (Source: Takeuchi, Kazuhiko (ed.), 1995)

(1) Increasing income (living standards)
(2) Higher consumption
(3) Effective use of resources (energy efficiency, recycling, etc.)
(4) Availability of resources
(5) Maintenance and nurturing of the natural environment
(6) Provision of amenities (leisure, comfort, contact with nature, etc.)

who will use the designed system is often influenced by the socio-economic and political structure of the society and its standard of living. It is these levels of needs and use that initially determine the size and extent of the pattern of the built form and the servicing systems (see Fig. 21). It is therefore ridiculous to assume that some internal environmental conditions are universally applicable. Studies have shown that people living in hot regions, especially in developing countries, usually accept higher temperatures and/or humidity levels because of lower expectations and natural acclimatisation.

In European low-energy social housing studies for temporate climatic zones, the fuel energy consumption is about 125 kilowatt hours per square metre (kWh/m^2) per annum in residential developments for 2 occupants per dwelling, and 166 kWh/m^2 per annum for 3 or 4 occupants per dwelling (in Desmecht, et al., 1998). The impact and/or damage to the environment by human beings is directly proportional to their living conditions and standard of living. Traditional, pre-industrial societies existing at a subsistence level obviously make far fewer demands on their environment (the current fascination with such societies in the developing world and the interest in such ideas as the Native American one of respecting and protecting nature is a phenomenon which unfortunately cannot be gone into here). But once human activities, modes of existence and expectations go beyond those of a 'simple existence' (e.g., when they demand a greater or managed food supply, heated and cooled shelters, more developed locomotion, etc.), environmental impact starts to be more noticeable. Indeed, as examples of rapid and unmanaged industrialisation show, it can become catastrophic. The more people depart from a simple way of life, the more demanding and complex are their interactions with the environment, from which they draw (or drain) more and more support and resources. As a result, they will have to expect and anticipate more ecosystem impairment.

From the planner's side of this equation, it can be seen that a reduction in human environmental influence is possible, but only at the expense of a reduction in the provisions for shelter and comfort. The less demanded by people from ecosystems, the less will be the human impact on them. If people had no need for shelter and comfort, then there would be no necessity for an ecological approach to the design of the built environment because human beings would be a completely integrated part of nature. For instance, with the use of digital technology, many of our current processes can be done electronically, in which case there will no longer be a need for many of the structures for direct human interaction. In an already industrialised world and in a society with already high material expectations, the designer has a difficult task in balancing comfort levels and energy use with environmental impact. For example, if a skyscraper is to be built in the United States, the designer will note that the typical American uses 11.5 kW of energy per day (in Steele, op. cit.), and the designer would therefore need to evaluate whether the existing projected energy consumption rates of the users for skyscraper in this locality might be reduced.

Users also affect the extent of waste and discharge from the building. This requires that the designer be familiar with discharge figures and patterns for waste.

Organic food waste is the highest, followed by paper/cardboard and plastic/rubber waste. Waste paper production in commercial offices (figures for the United States) is about 0.110 kg/sqm per day. This being the case, we can either ensure that appropriate provisions for recycling or reuse are in place, or alternatively we can seek to influence by design the behaviour of the users in the locality by ensuring that they reduce the extent of waste output. For example, economical urban waste production is regarded to be about 0.8 kilograms per person per day, whereas 'spendthrift' waste production is about 1.5 kilograms per person per day (in Penportier et al., 1998). Savings can also be achieved in lowering cold water use from 150 litres per person per day to 80 litres per person per day.

Influencing the user's needs here can be a simple management decision, such as just requiring the male users to remove their necktie 'which can reduce about US$50 per month worth of air-conditioning and power-supply equipment' in the United States (Von Weizsacker et al., 1997). For another example, if occupants of a non-domestic building treated that building with the same attitude that they have to their homes, energy management would be simplified. In winter, people are often willing to wear more clothes at home to reduce the impact of cold weather on heating bills. There is no reason why they should not do the same at work. A reduction of 1.25 °C in heating temperature can have a significant effect on lowering energy consumption, and can be achieved by encouraging a sensible attitude to seasonal and weather-related dressing. Similarly, in summer, light clothing and natural ventilation (i.e., opening windows) is preferable to mechanical ventilation or air conditioning on both health grounds and in terms of energy impact. Some of these decisions are management decisions for the end-users of the building, but obviously if the building has been designed with windows that do not open certain options are foreclosed.

Buildings – permanent shelters and workspaces – are now considered pre-requisites for human existence. Hence the problem for architects is that, since such structures are going to be made and used anyway, how can they be constructed in such a way as to have as little effect as possible on terrestrial ecosystems and natural resources? But the extent of environmental impact of a structure reflects the values of the society that found a need for the building in the first place. For although it is a fundamental characteristic of all life to take in suitable materials (e.g., food and air) and convert them into products of value to its own or its species' survival (e.g., heat and metabolic energy), in contemporary human society the intake (i.e. inputs) includes materials such as fossil fuels to satisfy complex needs for energy, shelter and waste disposal (Detwyler and Marcus, 1972), and the extent of this intake (and hence the related wastes) is tied to the standard of living (or gross national produce) of the people in the society where the project site exists. It is an inescapable fact that in order to provide people with the intakes required by their mode of existence, inevitable changes to the ecosystem will be incurred. Therefore, before finalising the design brief (which determines the extent of construction and enclosure), a questioning of the project programme and the users' need of that locality is beneficial. We might find, for instance, that these needs might be met by other means without the provision of any enclosure at all, or even by means of partial enclosures or by simply seeking a consensus among the users to reduce their level of resource intake and environmental requirements.

As mentioned above, the initial impact of a design on the environment is proportionately related to the size and context of the user's design requirements. We can regard the preparing of the design brief as the setting down of standards of comfort (both spatial and environmental) for the users of the designed system (e.g., acceptable internal designed temperature) and of consumption. Having established the extent of our designed system (i.e. its gross area), we next need to determine the extent to which we can lower internal environmental standards for the structure (e.g., the internal design temperature, air change/hour, indoor air quality and humidity, etc.). These levels will depend on the extent to which the people who will use the built environment will accept them. The higher the level of perceived need, the more extensive will be the size of the built environment, and therefore the greater its ecological impact. For instance, if a lower working internal temperature is acceptable to the users of apartment or office buildings, then the building's M&E systems can be designed operationally to consume less energy. Another energy efficiency strategy might be to limit the design temperature of the internal spaces to minimise winter and maximise summer temperatures (e.g., to 19.25 °C and 25.25 °C, respectively).

Setting these rates is no easy task. At what point does it matter that a space is 'too hot' or 'too cold'? More significantly still, whose decision is this to be? When does 'discomfort' become a real problem, when is it a problem worth doing something about and what environmental costs are we prepared to pay to resolve it? As I have mentioned, people used to living in hot regions, especially in developing countries, usually accept higher temperatures or humidity levels, not only because of natural acclimatisation but also due to lower expectations. Thus, such questions bring up larger lifestyle and cultural issues.

The size and extent of spatial accommodation and the quality of environmental comfort provided in the designed system affect not only the ambient ecosystem of the project but also the quantities of energy and materials (the earth's resources) that are consumed and depleted. As these obviously depend on the standard of needs and use required by the people who will use the built environment or the people who commissioned its design, then it is clear that the higher the levels of needs and use, the more extensive will be the size of the designed system and the extent of its sub-systems (and operational systems), and consequently the more will be its resource demands and ecological impact. If the designer is taking a green approach and is trying to lower levels of consumption by setting lower values on comfort and energy levels than those prevailing in the locality – in short, if the designer's building is going to be more ecological and less spendthrift than the buildings around it – then the end-users of the building will have to tolerate the designed changes in living comfort and consumption conditions. The ecologically minded designer has to deal with these issues in advance of preparing a design brief in terms of professional liability, user education, etc. which are outside the scope of this book.

In evaluating the user's needs, consideration needs to be given not only to internal environmental standards, but also to the consistency of provision of internal environmental standards. If passive mode environmental systems, i.e., non-use of any electro-mechanical devices or operational systems, are adopted, the consistency of internal conditions may be variable since passive mode systems are ambient-climate dependent. This becomes a subjective issue, dependent upon the individual preferences of the users of the buildings themselves. It might be argued that it is in the preliminary design stage that the architect is in a position to have the greatest influence on the extent of impact of the built structure. In terms of energy inputs, half of all energy used in a nation (Edwards, 1996, p. xiv) is in the heating, lighting, cooling and ventilation of buildings. In terms of outputs, the way buildings are designed, serviced and adapted over time all directly influences the volume of fossil fuels consumed, which relates directly to the volume of carbon dioxide released into the atmosphere, a factor in the raising of planetary temperatures (the greenhouse effect).

At the beginning of the production of the design brief, it should be ascertained whether it is possible to meet the skyscraper's comfort requirements largely through a design incorporating passive mode measures with a direct effect. In any event, the design strategy must begin by optimising all the passive mode strategies (see chapter 7). Following which, the designer must endeavour to use those mixed-mode systems that are viable and acceptable. The remaining energy needs in terms of heating, cooling, electricity and ventilation should be met by those active systems powered by ecologically sustainable forms of energy.

The provision of the internal operational M&E environmental systems might be reduced if greater attention is given to the use of the ambient energies of the location through bioclimatic design principles; users' needs can be reduced by a climate-responsive building configuration, or by passive devices or appropriate building orientation, rather than through the building's hardware as full-mode strategies. To achieve this goal, the designer has to be prepared to 'engineer' the architecture of the skyscraper and other intensive building types, their configuration, orientation, external-wall design, M&E systems and other characteristics with careful consideration to the climatic features of the project site. These features include the solar path, wind pattern, humidity and other factors (passive design is further discussed in chapter 6).

Passive-mode designs do not preclude using composite systems that also have mixed-mode and active-mode devices. For example, creating a climate-responsive building configuration requires the designer to respond inventively to the climate and ecology of the location (including the latitude and the ecosystem) through siting, orientation, layout and construction; but the architect must also selectively design the building's M&E systems for their contribution to energy consumption and conservation after first using the natural ambient climatic energies to their fullest. Of course, such an 'engineered' design solution should not in any way inhibit creative interpretation. The design needs to be evaluated quantitatively and to be based on a rationale. We need to be aware that the ecological approach is not a set of regimented design rules that result in a deterministic set of built forms. Variations are possible, and inventive means of compensation for deviations from the norms can be adopted.

It is important not only to re-evaluate user requirements – the patterns of use in the building are important as well. For example, U.S. businesses alone consume an estimated 21 million tons of office paper every year – the equivalent of more than 350 million trees. In fact, office paper is one of the top six contributors to waste outputs from office skyscrapers and among the fastest growing by percentage. Now, if the pattern of use in offices throughout the United States increased the rate of double-sided photocopying (e.g., through user education), it would save the equivalent of about 15 million trees (Goldbeck, 1997, p. 271).

It will be useful to categorise the level of operational systems provided to our skyscraper and other intensive building types, in other words the extent of its internal environmental servicing systems (dealt with in chapter 7) into three levels of provision (the following adapted from Worthington, J., 1997, p. 11):

- passive mode
- mixed mode
- full mode (or specialised mode)
- productive mode

The provision of the basic level of systems at the passive-mode level, if acceptable to all occupants, is ecologically ideal. It requires the optimisation of all possible passive-mode systems for the locality (e.g. Vale, B., and Vale, R., 1991). The full conventional systems level of servicing is referred to here as the specialised level or the full mode. The in-between or mixed-mode level is the background level of servicing. Productive mode is the use of systems that generate energy (e.g., photovoltaics). The designer must decide at the onset which of these levels of operational systems is to be provided in the building.

We can conclude that in the ecological approach, the designer must start with the premise that the environmental impact increases in relation to the increase in demands by users for living conditions beyond those of a simple existence. The first question to be asked prior to design is, "what is to be built?" and to assess its validity and consequences generically. In preparing the design brief, the designer must find out the extent of shelter and comfort that he or she must design for.

Philosophically, it is important to note the ultimate implications of green design. If the designer were to attempt to keep all adverse environmental impacts of a design to an absolute minimum, then it could mean that society may well have to return to a much simpler form of existence (i.e. lower than the current general standard of living), as well as to living conditions that make fewer demands for environmental comfort, shelter, energy and materials consumption than the present ones do. However, this may require a complex and extensive restructuring of the existing sociological, economic, educational and political structures, which is obviously outside the realm of the designer.

By considering the ecological design holistically in terms of the four factors in the partitioned matrix, it is clear that ecological design must encompass not just architectural design, engineering design and the science of ecology but also other aspects of environmental control and protection such as resource conservation, recycling practices and technology, pollution control, energy embodiment research, ecological landscape planning, applied ecology, climatology, etc. The partitioned matrix here demonstrates the interconnectivity of this multitude of disciplines which must be integrated into a single approach to ecological design.

Chapter Five

The Environmental Context for Building

The City's Hinterland and its Ecological Context

Having defined the requirements of the building to be designed, the next step is to look at the project site as an ecosystem. All ecological design must examine critically the ecological and climate characteristics (and the natural boundaries) of the project site. We can further assert that any design that does not take into consideration the ecological and climatic characteristics of the project site and how they impinge on the designed system and its operation cannot be considered ecological. This point must be emphasized because many designers totally negate the site's ecosystem and focus their design efforts entirely on the building without realising that their building exists within the biosphere and has to interact with it. The project site does not exist in isolation but is ecologically connected to the ecosystems of the locality and to the natural processes in the biosphere, all of which must be taken into account.

Our design objective is to achieve a symbiotic relationship between our manmade system and the ecosystem and to make the landscape (ecosystem) surrounding the designed system into an intrinsic seamless part of its architecture, that is, to create 'building-as-landscape' and 'landscape-as-building'. To achieve this goal, the designer must take into account the external ecological interdependencies of the designed system (defined by L22 in the partitioned matrix described in chapter 3) and how they may be incorporated in the design process. The external ecological and physical context of the proposed designed system includes the ecology of the urban site as well as the ecological systems in the hinterland of the entire urban environment. Within the biosphere are bio-geographic provinces, which are areas of land that share similar vegetation and climate. The natural resources available in a given location (e.g., sun, wind, etc.) should be optimised for the passive operations of the building, which are reflected in the layout and the built form.

Green design requires that all the systems and activities of the proposed building be evaluated, at a point early in the design process, for their potential to change and/or damage the ecosystem.

Building construction, like other human activities, always produces some kind of change in the environment; this modification can range from very destructive to beneficial, but any change is essentially an environmental impact. Every activity has the potential to be detrimental or contributive. The degree of the effect of a specific action will depend on a variety of factors, including: the density and intensity of the activity; its duration; the strength or carrying capacity of the ecosystem involved; and the presence of other activities in the same area which can either moderate or accentuate the effect under examination.

For the purposes of building design, the useful life of the building represents the duration of the 'activity' (i.e. creating and maintaining a built environment within the natural one). The potential harm represented by the design and the tolerance level of the environment at the site combine to determine the environmental impact. In some cases, an ecosystem might be able to tolerate a certain effect over a short period, and this can be factored into the activity – i.e. the activity can be limited so that it does not overstress the ecosystem's capacity to withstand the changes inflicted on it.

As ecological studies have shown, since one ecosystem is connected to other ecosystems within the biosphere, an action on the ecosystem of the project site (which may not have immediately apparent impact at that location) may have detrimental impacts on ecosystems elsewhere. The importance of the impact of any human action on the ecology of the project site will depend on the ecological condition and value of the local ecosystem as well as on the type of action that is to be pursued.

The designer must therefore examine the ecosystem where his project site is located before imposing any action upon it.

Discussed below are some of the methods and indicators that may be used in ecosystem analysis as part of the designer's investigations of the project site and its surroundings. Generally, we will find that an ecosystem analysis of the project site will provide the designer with the basis for determining the type of land use, preservation areas, conservation areas, the siting and built-form patterns and likely impacts during the life cycle of the designed system.

Hierarchy of Ecosystems

We can classify the project site into one of the following six categories:

● Ecologically Mature Ecosystems
These ecosystems have very high biodiversity. Ecologically mature
ecosystems include forests, deserts, wetlands and rainforests and
essentially ecosystems that are not directly affected by any manmade interference.

● Ecologically Immature Ecosystems
These are ecosystems that, while still natural, are recovering from
damage or are in the process of succession or regeneration.

● Ecologically Simplified Ecosystems
These are sites that, though originally mature or immature, have
now been savaged by grazing or controlled burning, by being mown
or selectively logged, and by the removal of biotic components.

Ecosystem hierarchy	Site data requirements	Design strategy
Ecologically mature	Complete ecosystem analysis mapping	• Preserve • Conserve • Develop only in no-impact areas
Ecologically immature	Complete ecosystem analysis mapping	• Preserve • Conserve • Develop in least-impact areas
Ecologically simplified	Complete ecosystem analysis mapping	• Preserve • Conserve • Increase biodiversity • Develop in low-impact areas
Mixed-artificial	Partial ecosystem analysis and mapping	• Increase biodiversity • Conserve • Develop in low-impact areas
Monoculture	Partial ecosystem analysis and mapping	• Increase biodiversity • Develop in areas of non-productive potential • Rehabilitate the ecosystem
Zeroculture	Mapping of remaining ecosystem components (e.g., hydrology, remaining trees, etc.)	• Increase biodiversity and organic mass • Rehabilitate the ecosystem

Fig. 22   Site ecosystem classification (Source: Yeang)

● Mixed Artificial Ecosystems

These are mixed ecosystems which are artificially maintained by man, e.g. through crop rotation, agro-forestry, parks, gardens, etc.

● Monoculture Ecosystems

These are again artificial, but monocultural, ecosystems (e.g. agricultural use, replanted forests for timber harvesting, plantations, crops, lawns).

● Zeroculture Ecosystems

These are totally artificial ecosystem sites with zero remaining ecological culture, e.g. urban sites, open-cut mines.

Going down the hierarchy, the biodiversity decreases, as does the degree of natural (ecological) control of processes. At the same time, the demand for energy and maintenance input (e.g. addition of fertilisers, weeding, etc.) increases and the fragility increases.

At the start, the designer determines which category the project site belongs to and the extent of ecosystem analysis and mapping to be carried out (see Fig. 22).

Ecosystem Modelling and Mapping

The sustainable approach to site planning and design goes beyond combining and comparing site inventories, and attempts to determine the relationships between site elements and how the elements will adapt to change. Understanding these relationships also clarifies how development impacts from one area of the site will affect other areas. For example, we will frequently find that it is a location's hydrology that has been a major determinant of the shape of the land and soils, and correspondingly, as we might expect, of the plants and wildlife. By measuring water quality within watersheds, a good understanding of the functioning of the system can be derived.

An evaluation of potential development impacts requires that a predevelopment baseline or environmental model for the project site be produced. This model will describe the essential functions and interrelationships of the individual site factors and will establish acceptable limits of change during and after construction. Selected environmental monitoring and testing would also be done during construction. The entire building of the development might be phased to allow time between construction periods to monitor environmental impacts and adjust the baseline model.

The major steps in a sustainable approach to site planning and design are as follows:

- Model the ecosystem of the project site to establish an understanding of the environment.
- Establish acceptable limits of change to protect those parts of the natural environment that must not be developed and should be buffered, either because they are rare, ecologically fragile or both.
- Design the built systems within environmental thresholds and to reintroduce natural features into the developed areas.
- Monitor site factors throughout construction and operations.
- Re-evaluate design solutions between development phases.
- Restore those natural systems and areas that have previously been damaged by human activities.
- Productively contribute to the site's ecology by increasing the biomass of the inorganic areas of the building.

In addition there would be the usual social and economic considerations of the intended development or building and the content.

The extent of ecosystem modelling and mapping depends on the project site status in the earlier hierarchy of ecosystems. Obviously, the ecologically mature project site requires the most extensive analysis of its ecosystem and understanding of the effect of manmade changes. This analysis decreases in complexity until we reach the zeroculture ecosystem site, from which most biotic components have already been removed.

It is not only decisions about the scale and content of

In aggregate, the urban built environment, with its complex matrix of buildings, activities, services and transportation, consumes 75 percent of the world's energy resources and produces the vast bulk of its pollution and climate-changing gases.

our buildings and towns that are important; we should also examine their spatial distribution (i.e. their physical planning) in the ecosystem, which is crucial to creating a sustainable future. Spatial displacement through building physically destroys ecosystems.

There are further, less obvious but related costs. Buildings are responsible for about 50 percent of total energy used in a country, and the transport needed to get to these buildings and the provision of supplies to them from rural and outlying areas account for another half of the remaining energy consumption. Thus it is clear that ecological design must also give consideration to the transport routes and patterns of access to any proposed building and the energy consequences of people coming to and from the building, especially during its operational phase.

At the large-scale level, the external ecological interdependencies of the built environment consist of the totality of all the earth's ecosystems and its resources. An awareness of the features of the ecosystems on which human activities are to be imposed provides baseline criteria for the management of the major spatial displacement brought about by building. Ecosystems provide diversity between and among biotic communities. Natural systems, resources and reserves should be protected and managed for their long-term well being (and ours). Some natural environments should not be developed because of their extreme fragility. Designers can play a role in the safeguarding of natural places. Areas to protect include forest, storm water storage areas, natural springs, rivers, streams, coastal waters and shorelines, recharge areas for aquifers, areas of high seasonal water table, mature vegetation, steep slopes, and wildlife and marine habitats.

Let us consider a few of these ecological features in detail. The great value placed on watersheds comes from their ability to absorb and cleanse water, recycle excess nutrients, hold soil in place and prevent flooding. When plant cover is removed or disturbed, water and wind can race across the land taking valuable topsoil with them, and it is a fact that soil, once exposed, is eroded at several thousand times the natural rate. Under normal conditions, each hectare of land loses somewhere between 0.04 and 0.05 tons of soil to erosion each year – far less than what is replaced by natural soil building processes. On lands that have been logged or converted to crops and grazing, however, erosion rates are many thousands of times higher than that. The eroded soil carries nutrients, sediments and chemicals valuable to the system it leaves but often harmful to the ultimate destination. As a general strategy, where the impact of a designed system or an intended activity on the ecosystem could cause detrimental changes, the implementation of the action should be weighed with the preventive or corrective measures that could be incorporated into the design, as well as possible alternative design solutions. To do this adequately, a thorough understanding of the local ecosystem is crucial. The ecosystem can be described in a number of ways. A method commonly used in land use planning is the 'layer-cake' method. Other methods include studies of the energy flows through the ecosystem (energetics).

The impacts of all the interdependencies of the built environment on the earth's ecology and resources (L22 in the partitioned matrix) are vital considerations in the design process. Actions at the building site can have profound effects. Deforestation (for timber harvesting and for urban development) contributes 15 percent to global warning. The provision of inputs in the creation of a built system, the emission of outputs from a built system and the operational activities within a built system after it is constructed all have effects on the components and the airborne and waterborne processes of the ecosystem. Impacts must be considered over the lifecycle of the designed system and should include local, regional and global effects. Clearly, an awareness of these effects at the onset of the design process would facilitate future computation of the ecological consequences of other intended built structures, as well as providing a basis for minimising undesirable future changes to the ecological environment of the locality of the project site. Observation of the site will reveal to the designer important physical features, such as the pattern of natural drainage, significant land forms, vegetation and climatic conditions. In the case of urban centres, where most skyscrapers and similarly intensive buildings are located, we need also to mitigate the urban 'heat island' effect by increasing the volume of organic matter in the urban environment, e.g., through landscaping. Some of the ways to do this are by selective distribution of vegetation, the revision of the layout of city blocks, choice of building configuration (size and clustering), the colour of roofs, surfaces and buildings, and the properties of surface materials (e.g., their reflectance, heat absorption and build-up consequences, etc.).

We should acknowledge that these external interdependencies (L22) include global interdependencies of ecosystem processes and resources. The built environment is dependent on its external environment as a supplier of energy and material resources for its physical form and substance, as well as for maintenance of its operations. If the long-term supply of these resources is to be ensured, then a conservationist approach to their use should be a design criterion. Immediate design objectives should be to provide the potential for and flexibility in future resource use. The ecosystems of the project site can also serve as a sink to assimilate the discharges from the designed system (see chapter 3). As we have seen, the capacity for assimilation by the ecosystem is limited and if the threshold is exceeded, the ecosystem will deteriorate.

The Ecological 'Sieve-Map' Method

Especially for ecologically mature project sites, the designer needs to carry out an ecological analysis. The extent of its complexity obviously depends on the ecological history of the site. We need to appreciate the functioning of the ecosystem and establish rules of conduct for the people who use it. One of the more common and easier methods of doing this is the ecological land-use planning technique known as the sieve-map method (McHarg, 1969).

This method involves the simplified mapping of the ecosystem in terms of its physical natural features (vegetation, soils, groundwater, natural drainage patterns, topography, hydrology, geology and so on) and through the evaluative technique of 'sieve mapping',

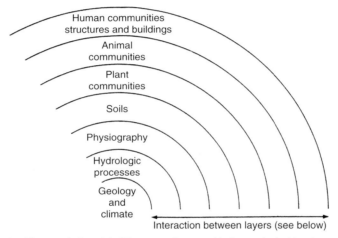

A structural 'layer-cake' model of the ecosystem used in the ecological land-use planning method

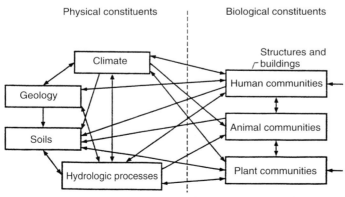

Interactions between the layers (i.e. the physical constituents and biological constituents) from the 'layer-cake' model. Many of the linkages are reciprocal

Fig. 23   Interactions within the ecosystem (Source: Yeang, 1995)

which produces a plan showing land areas suitable for different intensities of development and building types in relation to the carrying capacities of the natural systems (see Fig. 23). In the site planning or master planning of built systems on large sites, the buildings and roads can then be located at appropriate points to ensure the least ecological impact and minimal interference with the ecosystem.

The production of the ecological analysis or profile of the site of the proposed building can be a daunting task, requiring significant time and expense. Because an ecosystem goes through natural seasonal changes, a complete analysis would have to include observation of the site at various times of the year – and the more diverse the site is (i.e. the higher up on the scale of hierarchies) the more complex is the task. The flora and fauna of the building locality provide the designer with clues to the condition of the site's ecology. As has been pointed out, a total ecological analysis including all interrelationships is a huge task, so the use of the condition of biotic communities as a benchmark is a helpful way to estimate the overall state of the local environment (Kaiser et al., 1974). An ecologist can use information about the status of plant populations to make inferences about the status of the site's soil, micro-climate, hydrology and animal species. For example, a particular growth of vegetation and its state of succession tells much about how recent has been any ecological disturbance of the area, the biological productivity of the site, and the biodiversity and stability of the environment. Further inferences could be drawn about the interaction between the local ecosystem and other surrounding environments.

By such careful observance of the site ecology, the designer can forecast the degree to which the planned built system is suitable to the locality and the ecosystem's vulnerability to damage from human intervention such as building. The 'layer-cake' method mentioned above can be simplified into a series of steps, as follows:
- Identification of biotic (faunal and floral) species present, including their diversity, distribution over the site and numbers.
- Establish the relationship of species populations to physical and biological processes.
- Create a hierarchy of species, i.e. determine which are most valuable and crucial to the functioning and viability of the ecosystem.
- Factor these conclusions into the design and building plan so as to minimise changes to biotic communities and permanent alterations in terrain, including other factors such as edaphic factors, etc., and land configurations (see Figs. 24 and 25).

The description of these biotic and abiotic components of the site's ecosystem, it should be remembered, only serves to forecast the

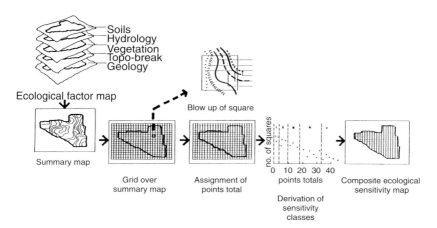

Soils
Hydrology
Vegetation
Topo-break
Geology

Ecological factor map

Summary map

Blow up of square

Grid over
summary map

Assignment of
points total

no. of squares

0 10 20 30 40
points totals

Derivation of
sensitivity
classes

Composite ecological
sensitivity map

Fig. 24   Ecological land-
use planning; sieve map
technique method (Source:
Takeuchi, 1995)

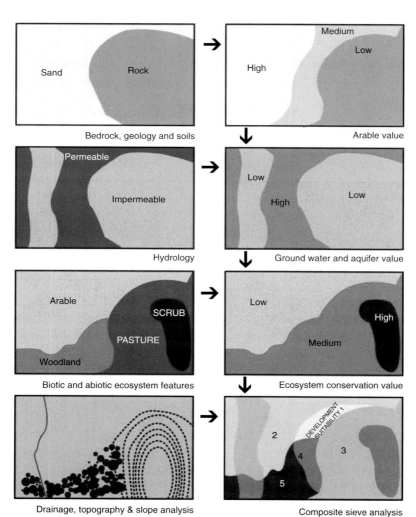

Sand   Rock

Bedrock, geology and soils

Medium
Low
High

Arable value

Permeable
Impermeable

Hydrology

Low
High   Low

Ground water and aquifer value

Arable
SCRUB
PASTURE
Woodland

Biotic and abiotic ecosystem features

Low
High
Medium

Ecosystem conservation value

Drainage, topography & slope analysis

DEVELOPMENT SUITABILITY 1
2
4
3
5

Composite sieve analysis

other functions and systems of the environment, and is not an exhaustive ecosystem map. A total analysis would have to look at the other sub-systems as well and establish the links and interrelationships between them and the plant and animal populations.

Ecological elements can be broken down into several categories. Plant and animal species are one. Other factors are habitats and ecological processes. These elements can in turn be further specified. When the designer looks at the species present in a site, he or she must consider whether rare, threatened or endangered plants or animals are present; game animals; migratory birds or colonies; and pestilent species (plant or animal) and/or parasites. A consideration of habitats will involve determining food chains, diversity of species and appropriate land use. The processes of the ecosystem include its productivity, hydrology and nutrient rates. There is disagreement as to how and even if one should prioritise these elements. Ultimately, this will be the designer's decision, for not even biologists are agreed about how to protect an ecosystem from damage by humans. The approach may be total, or one might focus on preserving certain species in one's use of the site for development. The decision as to how and how much of the ecosystem will be conserved has to be taken in the context of the particular site and its unique qualities, as well as the designer's views and goals.

One of the elements mentioned above, biodiversity, is a useful method for comparing impacts (e.g., in Ismail, G., and Mohammed, M., 1997); indeed, it is often held to be the broad foundation for assessing sustainability. Biological diversity means, 'the variability

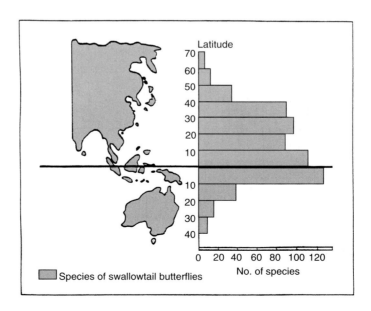

Fig. 25 Species diversity in relation to latitude (Source: Cranbrook, E., and Edwards, D. S., 1994)

Group	Described species	Estimated species	Percentage of total already described
Micro-organisms			
Bacteria	3,000	2,500,000	0.1
Plants			
Algae	40,000	350,000	11
Bryophytes	17,000	25,000	68
Vascular plants	220,000	270,000	81
Funghi			
(including lichens)	69,000	1,500,000	5
Animals			
Nematodes	15,000	500,000	3
Arthropods	80,000	6,000,000	13
Fish	22,500	35,000	64
Birds	9,040	9,100	99
Mammals	4,000	4,020	> 99

Fig. 26 Estimates of global biodiversity and alpha diversity (Source: Cranbrook and Edwards, 1994)

among living organisms from all sources including inter alia, terrestial, marine and other aquatic ecosystems and the ecological complexes of which they are part; this includes diversity within species, between species and of ecosystems' (Spellerberg, 1996). Simply stated, this term describes the diversity of all living things including plants, animals and micro-organisms (species). It also refers to the diversity of their living places (habitats) and of the functions performed in those places (niches). The concept also relates to the diversity of genetic material (genes) carried by the organisms. Each of these aspects of biodiversity is important for securing a sustainable future for the environment and for human society (see Fig. 26).

The species diversity of a stated group at a particular place and time is termed **alpha diversity**. All species richness of a sample may be expressed by a diversity index which is dependent on sample size.

One such index is Williams's alpha, derived from the equation:

$$a = S/\log_e (1+N/a)$$

where N is the number of individuals and S the number of species in the sample.

Biodiversity has four levels:
- diversity between and within ecosystems and habitats
- diversity of species
- diversity of niches
- genetic variations within individual species

Species Diversity
The number of individual species in an area is sometimes called species richness. Ecosystems vary in richness. Examples of species-rich ecosystems are rainforests and coral reefs; species-poor ecosystems are alpine areas, deserts and most designed environments.

There are also variations in how evenly or equitably the numbers of individuals within species are distributed. For example, there could be small numbers of many species as in a tropical forest, or large numbers of a few species as found in a eucalyptus forest, estuary or designed landscape.

There is no simple rule of thumb for the relationship of species diversity to stability. However, species loss can be catastrophic for both species-poor and species-rich ecosystems.

Habitat Diversity
This term describes the number of different physical environments, or micro-climates, each with its own range of organisms, that occur within an ecosystem. Examples are snowfields and stream margins in an alpine area, treetops and forest floor in a eucalyptus forest. The overall biodiversity of an area can be a result of great habitat diversity. Examples of low habitat diversity are grasslands, tidal flats, oceans and most designed landscapes. Habitat loss can have a more significant effect on biodiversity than species loss, because new species can migrate into an area but habitats cannot repair themselves. Habitat preservation is more important than species conservation.

Niche Diversity
This element describes the variety of relationships that can occur between organisms and their habitat. Species specialise and adapt to specific 'occupations', which may involve exclusive relationships with other species, and avoid competition for space and resources. For example, there are plants that can be pollinated by both bats and birds; bats and birds avoid competition by having different nesting sites but visit the same plant species. Some habitats have high niche differentiation (e.g. eucalyptus forests, semi-arid areas and coastal scrubs).

Niche diversity is related to both species and habitat diversity. Loss of habitat results in decreased niche diversity, which is evidenced by loss of species diversity. Grazing and mowing the undergrowth of forests reduces niche and species diversity. Niches can then be occupied by less well-adapted species, such as feral animals and plants, which in turn may displace more indigenous species and further destabilise the ecosystem.

Genetic Diversity
Commonly described as the 'gene pool', genetic diversity refers to the totality of all the genes in an ecosystem. It is a rather abstract term; however, it has the desirable quality of making us aware of

the necessity to conserve and protect the information of the ecosystem and to realise that each species is precious, even spiders, fungi and bacteria, for the irreplaceable genetic resource they contain.

It can be argued that it is important to maintain the diversity and abundance of plant life even if it is not required immediately for human needs. Only 12,000 of the 220,000 flowering plants that grow on the planet are used for human needs, and of these only 150 are grown commercially. Of those 150, however, 3 species (wheat, rice and maize) provide 60 percent of our global food requirements. The species diversity of a stated group at a particular place and time is termed 'alpha diversity'. When all the species cannot be accommodated for practical reasons, the species richness of a sample may be expressed by a diversity index, which is independent of sample size. One such index is Williams's alpha diversity index (see Fig. 26). The designer must ensure that his actions do not result in a decrease in biodiversity and that all possible opportunities to restore and increase biodiversity are taken. Some of these actions are as follows:

Activities that reduce biodiversity	Actions that increase biodiversity
Site works	
● Clearing	Limit and restrict cleared areas, see uncleared area as a resource
● Grading	Accept greater limitations, retain & stockpile materials
● Tidying	Accept and design for greater diversity, resist interference
● Draining	Accept higher constraints, design for on-site water use
● Paving	Reduce hard surfaces to minimum, use porous surfaces
● Excavating	Accept slope constraints, excavate only when essential
Building process	
● Waste	Provide site sanitation; restrict waste to specific locations
● Pollution (lime, cement)	Confine mixing and wash sites, regenerate on completion
● Compaction	Reduce and restrict heavy activities, regenerate sites
● Parking	Confine worker's parking to off-site

● Contamination	Confine on-site movements to designated areas; regenerate
● Indiscriminate damage	Supervise for minimum disturbance, fence off restricted areas

Building design features

● Excessive site coverage	Reduce footprint, consider building upward, reduce internal circulation space, level space for wildlife movements.
● Excessive bulk	Increase external circulation space, consider integrating indoor/outdoor spaces with habitats for wildlife
● Overshadowing	Orient and design to give sun to surrounding areas as well as buildings
● Wind effects	Design to avoid excessive wind speeds and turbulence in natural areas; don't make wildlife corridors into wind tunnels.
● Location of glazing	Site glazing so as to avoid migration routes and flight paths of birds; evaluate such paths as part of site appraisal
● Obstruction of movement	Site and design to allow natural movement of animals, plants and processes such as shifting sand; allow natural waterways to pass under structures where practicable
● Excessive hard surfaces	Reduce hard surfaces to practical minimum, utilise porous surfaces, use run-off water on site to recreate habitats
● Animal-unfriendly surfaces	Use textured and natural materials where acceptable to provide habitats for insects, birds and lizards

Building and landscape materials and systems

● Polluting & space-intensive	Consider prefabrication of components off-site
● From industries which reduce biodiversity	Consider alternative materials/suppliers
● From locations where biodiversity is threatened	Consider alternative materials/ suppliers
● Toxic outputs or leachates	Use alternative materials
● Natural materials with short lifespans	Use more durable alternatives (provided they are from sustainable sources)

● From unsustainable sources	Use alternatives that preserve biodiversity
● Rare or threatened sources	Use only if available from sustainable sources, such as nurseries, or use more common species so biodiversity can recover
● From exploited locations	Consider paying more for materials from unexploited locations which can pay for the conservation of biodiversity

Landscape design

● Hard, formal schemes	Consider using soft materials and designing for habitat
● Excessive lawn and paving	Consider reducing monoculture and increasing diversity
● Water features	Consider creating water bodies with habitat and practical value; use water features to detain storm run-off
● Low species diversity	Consider schemes of higher species diversity and with a greater range of plant forms that can provide habitat diversity

Plant materials

● Rare or threatened species	Consider rare species so biodiversity can recover, provided they can be used sustainably
● From existing natural locations or nurseries	Use alternative materials or landscape materials from sustainable sources
● No habitat value	Consider using species with habitat value for indigenous wildlife
● Habitat for non-native species	Consider alternative indigenous species to provide refuge for native wildlife
● Potential pest species	Use alternative species that are not invasive pests and promote the use of these
● Non-indigenous species	Consider the use of alternative indigenous species so as to avoid the genetic pollution of surrounding areas by pollen

Landscape maintenance

● Mowing	Consider low-maintenance diverse landscapes instead
● Burning	Use appropriate-intensity, infrequent fires where possible and substitute hand removal of fuel from near buildings

● Herbicides/	Design for low maintenance and substitute
pesticides	labour for chemicals
● Fertilisers	Use natural products and composts where
	necessary
● Animals	Recommend against feral pets or
	design effective containment away from wild-
	life habitat

Site Planning Design Strategies

For the ecologically sensitive sites, after the architect has examined the environment and determined its biodiversity and other
factors, the next consideration is deciding how much disturbance
and displacement by the structure can be sustained by the ecosystem given its fragility and levels of tolerance. Attendant activities not part of the structure itself, such as the impact of transportation demands, also have to be considered, as well as the
built environment's operational systems inputs and outputs during its useful life.

The threshold of tolerance of the ecosystem is a key benchmark
for the designer because it establishes the level of impacts to
which the ecosystem is able to adapt and recover (as long as there
is sufficient time and provided that other impacts in the same area
do not further destabilise the environment). In making these calculations, however, the designer must keep in mind that it is easier to
protect an extant environment than to restore later one that has
been destroyed; species and populations, once exterminated, can
be difficult to re-establish in the ecosystem. The factors of resili-
ency, change and succession are interrelated (Holling and Goldberg,
1973). The development of ecosystems is evolutionary and follows a
series of stages of succession. The environment moves naturally
toward a stage of maturity, but passing from one level of the ecological hierarchies (dealt with above) to another is often not easy.
Obviously, it will be difficult for a zeroculture site such as an urban
site which has been completely paved over to be restored to the
condition of a mature ecosystem.

Conservation theory supports the idea that there are irreplaceable elements in a mature, biologically diverse ecosystem whose
importance must be seen in terms of their contribution to the stability of the biosphere over the long term (in Dasman, 1968). A particular ecosystem's elements provide the limits that our building
will have to observe, both as a physical intervention in the environment and as a built system with outputs that affect the ambient
ecosystem. The designer obviously has to know the limits of his or

her own design and all of its components and systems as well as the nature of the site. Only in this way can a 'match' be facilitated between design goals and what the ecosystem is able to tolerate. Some of the features of the ecosystem may be modified, while others of unique value may be placed off limits. Working within these constraints, the designer should address all the parameters before any actual intervention in the environment occurs – and when building does begin, it will have to be accompanied by adequate safeguards and monitoring to make certain that no crucial element of the ecosystem is permanently removed, harmed or interfered with without prior planning of conservation measures.

Ecological site planning seeks to enhance and protect the natural resources and biodiversity of the land. But construction of buildings is a process that requires an inevitable intervention in the environment. The environmentally conscious approach to site design takes into account a myriad of concerns and addresses them from the initial design conception, through assessment and inventory, to the construction process itself. The architect must also plan for the long term, examining what activities and systems will be operating during the life of the building. Each stage of the building's existence must be analysed in detail.

Questions have to be asked along the following lines: which environments are affected and what are their nature; will the ecosystem be physically, chemically and biologically altered; what flora and fauna occupy the projected site; what are the processes involved in the changes that can be anticipated from human intervention, i.e. building. Once the processes and the communities are known, the extent and type of environmental changes that will be wrought by the creation of the built structure can be anticipated.

As we have seen, permanent changes to environments are difficult if not impossible to amend. Heavy intervention also creates shock waves that extend local effects to other, interrelated ecosystems. Therefore, green design must aim for minimal impairment of the site's ecology. There are a number of strategies for doing this:

- Build on minimum site areas using small-footprint designs and leaving large portions of parcels undisturbed.
- Avoid excessive disturbance to soils and slopes to prevent landslides and structural failure.
- Avoid building on wetlands.
- Avoid clearing or grading steep slopes, especially those with sensitive geologic, groundwater and/or erosion factors.
- Avoid removing topsoil; sustain the site's percolation capacity and sewage drain fields; and protect the local groundwater supply, wildlife and marine habitats.
- Limit the volume and duration of surface water runoff by reducing the amount of land converted to impervious surfaces such as parking lots, roofs and roads. This reduces erosion, water degradation and flooding at lower elevations.
- Protect particular sites or places, such as productive arable land from being replaced by new building.

In the designer's planning of the built system's layout, the ecological analysis already carried out will give direction to a number of important design decisions that have to be made (after Wettgvst et al., 1971):
- Land-use patterns and the avoidance of environmentally unsuitable locations (including terrain that would be dangerous for buildings).
- The choice of areas of the site to single out for preservation because of their importance to the stability of the existing ecosystem. Identification of these areas may depend on biotic (i.e. fragile flora and fauna) and on abiotic (i.e. sensitive topographical features) elements.
- Selection of further areas which, though less critical for the ecology of the site, have some form of conservation value – for example as buffer zones between the preserved areas and areas of development.
- The planning of layouts on the site to accommodate both natural and manmade elements in a favourable pattern.

Areas that have already been developed and vacant land already in a condition to be developed are types of land that can be intensively used. Undeveloped land which may have problematic aspects such as drainage and percolation problems would be inappropriate. Areas suitable for building are usually those not considered ecologically fragile. Again, the designer has to look at impacts over time. The creation of the built environment can appear more or less invasive or damaging depending on how far into the future we project its effects. The accumulation or intensification of activities and the ecosystem's regeneration time and abilities are important factors in whether the built system's effect is negligible and/or temporary or permanently destructive. Negative effects could be broken down into several stages or levels. Some disturbances, such as an upstream dumping of damaging waste on a single occasion, will cause a temporary ecosystem change. A superficial alteration in topography could cause a corresponding surface change that disfigures the environment. And drastic intervention such as the complete removal of biotic components down to the bedrock in preparation for building would have permanent effects. Of course, a design can have beneficial impacts as well as negative ones. A green designer who tries to mate the built and natural environments in a positive configuration could actually preserve nature by creating reserves; add to the ecosystem's value by restoring abandoned or waste areas; reduce or reverse negative environmental trends such as erosion by incorporating manmade systems (in this case, drainage); and increase the ecosystem's biodiversity and viability by incorporating replanting or reforestation programs into the design.

In the selection and design of the built system's form and operational systems, the task is to seek compatible and positive relations between the operational aspects of the designed system and the project area's ecosystems processes. For example, the provision of buffer areas of trees and native vegetation can have a number of beneficial effects. They can serve as havens for resident and migratory insects and animals, which can pollinate crops and control pests. They may also help reduce wind erosion and control pollution that escapes from agricultural fields.

Some of the key landscape factors that should be considered and mapped in the ecological planning processes include:

Climate.   Specific climate characteristics should be considered in order to locate facilities for maximum human comfort and protection of the site resources and building facilities. The cooling effects, as well as the potential velocity and direction of prevailing winds, should be carefully reviewed. Also to be integrated into the site design is the heating, lighting and power-generating ability of the sun, along with the aesthetic effects of sunlight. Rainfall can be captured and drained in ways that will protect the soil; it can also be a source for primary use, which can then be recirculated for secondary use.

Topography.   In many areas, flatland is commonly set aside for agricultural use, leaving the remaining, primarily sloped lands for building. If handled properly, sloped topography can provide visual and sound separation. In site planning, the protecting of the vegetation and soil from erosion are primary concerns; the greater the slope, the faster the land will erode. Reducing the size of the building footprint, eliminating automobiles and parking provision and access roads, keeping soil disturbance to a minimum, elevating walkways and using point footings for structures are all potentially useful ways to protect the site topography. Integrating the existing geology of the site will also help to maintain its character and protect both the soil and vegetation. We should also avoid building in low-lying flood plain areas or on unstable soils, and should take protective measures to address adjacent water ecosystems and habitats.

Vegetation (flora).   To secure the integrity of a site, it is important to retain as much of the vegetation as possible. Sensitive local plant species need to be identified and protected. Existing vegetation should also be maintained to encourage biodiversity and to protect the nutrients held in the green canopy. Vegetation is significant because it protects wildlife habitats, nutrients, and soils; enhances visual beauty; offers acoustic and visual screening for privacy; provides a primary source of shade; and in some cases, creates opportunities for the production of food or other sustainable products. Protection and restoration of local plants is the fundamental purpose of ecological land-use planning in design.

Wildlife (fauna).   Sensitive wildlife habitat areas should be avoided. By maintaining as much habitat as possible, nearby wildlife is encouraged to remain close to human activity. Creating artificial habitats or feeding wildlife can have disruptive effects on the natural ecosystem. For instance, it has been calculated that 75 percent of the world's bird species are declining and nearly 25 percent of the 4,600 species of mammals are threatened with extinction.

Ecosystem capacity and density.   Every site's ecosystem has a carrying capacity for built development. A detailed ecosystem analysis should determine this capacity based on the sensitivity of site resources and the ability of the landscape's natural systems to regenerate. The siting of buildings should also carefully evaluate the merits of concentration versus dispersal patterns in layouts. Natural landscape features may be easier to maintain if the built systems are carefully dispersed; conversely, concentration of built structures (as in a high-rise tower) leaves more undisturbed natural areas.

Visual character.   Natural visual lines in the landscape should be incorporated wherever possible. Layout design should avoid creating on-site intrusions such as road cuts and infrastructural utilities. By working with the site's topography, locating small footprint buildings (such as towers) within existing vegetation, and working with the slope of the land, the visual aesthetics will reflect the natural and historic character of the locality. The existing landscape can be protected by careful coordination of construction methods and access around it, thus minimising collateral or inadvertent damage and saving on later landscaping costs.

Natural hazards.   Sustainable ecosystem habitats should be located with consideration for natural hazards such as precipitous topography and dangerous plants, animals, or water. The site layout should allow for controlled access routes to any dangerous areas.

Energy and infrastructural systems.   In any environment, there is the need for the built systems to receive running water, power and sanitary waste management. Early in the planning process, the designer must identify systems that will work within the parameters of climate, topography and natural resources of the site. For example, noise from mechanical equipment and treatment odors should be minimised and mitigated by location and buffering. Stabilising soils, preserving natural vegetation and capturing run-off in depressions are ways of regulating storm drainage in a revitalising manner.

Transportation and site access.   Access to the site for the purpose of construction, especially for large equipment, should be limited as much as possible to protect the existing vegetation and soils. Material staging and storage, as well as vehicular access and temporary power, must be considered prior to construction. The transportation routes and patterns should not break up the site into discrete islands. Their siting should also take into consideration their impact on the ecosystems.

Assessing existing toxins.   The designer must assess any potential toxins so that they may be addressed as needed. This means testing soil, water, and air for lead, radon, asbestos and mercury levels. In existing structures, lead and asbestos must be encapsulated or removed as appropriate.

On large sites, the orientation and planning of the built forms must be governed by the following ecological and bioclimatic factors (besides the ecosystem stability of the site's ecology):

- orientation of streets and building structures to the sun.
- temperature control and use of daylight in the public areas.
- topography (land form, overall exposure, general situation).
- direction and intensity of wind (alignment of streets, sheltered public spaces, systematic ventilation, cold-air corridors).
- vegetation and distribution of planted areas (oxygen production, dust consolidation, temperature balance, shading, windbreaks).
- hydro-geology (relationship to water and waterway systems).

Zeroculture Ecosystem Project Sites

The degree to which the ecological factors I have been discussing are of importance to the designer depends on the site's place in the hierarchy of ecosystems. Site location, geography, biodiversity and complexity will depend on the degree to which the area has already been opened to development and interference by humans.

The ecological approach will have obvious relevance to non-urban landscapes, where such matters as the risk of ecosystem damage, habitat destruction and displacement of agriculture will have to be carefully evaluated. Built systems threaten these environments more than some others.

The construction of a complex built system on a rural building site could cause habitat loss, elimination of certain species or promotion of competition between species, devastation of plant cover and/or plant species loss. Because of these hazards, stress has been laid on the need for a thorough ecological analysis at the very start.

The picture is rather different when we look at zeroculture sites. Here, so much human intervention has occurred over time and the urbanised space is so extensive already that biotic ecosystem components may have been completely eliminated. This leaves the designer with the task of preserving such ecological remnants as exist; however, it is of course necessary to still take into account factors such as climate, geology, soil conditions and hydrology. The skyscraper, through its foundations, could have an effect on the groundwater and other surface and sub-surface features of the urban locality. Other urban problems include possible shadowing of public or other adjacent areas by the skyscraper in different seasons (see Fig. 27). Particular urban flora and fauna may also exist.

In general, most existing urban and industrial environments where skyscrapers are likely to be located usually retain very little of the natural shape and structure of the original ecosystem. The previous biological structure and functional complexity of the ecosystems on which building development has taken place have often been replaced by a simpler, more synthetic and increasingly homogenized abiotic type of environment (e.g. concrete, asphalt, paving). With the creation of permanent urban settlements, the built environment has become more and more synthetic and intentionally remote from interaction with the ecosystems in the biosphere.

However, with a new structure, the new landscaping design may bring in new flora and fauna, which must be compatible with those which had been there before.

Transportation Considerations in Design

The production of a skyscraper results in the concentration of an increased population of occupants in a specific location. This concentration has obvious transportation consequences through people coming to and departing from the building, whether they are occupants, visitors or simply servicing personnel – and their vehicles (see Fig. 28). In addition, all ecological design and planning must take into account transportation within the site, as well as movement to the site. Transportation consumes about 23 percent of the energy used in a nation (see Fig. 29). Directly and indirectly, buildings contribute towards energy use through the transport implications of the form, type and location of development. For example, Europe in 1996 had 1 car per 2.5 persons (Edwards, op. cit.), making clear the importance of evaluating the transportation context of large buildings that may have hundreds

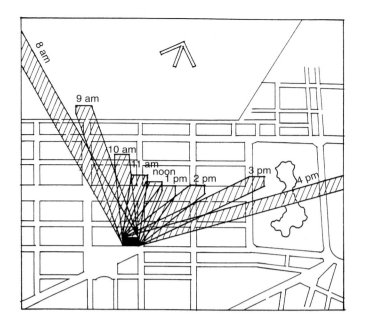

Fig. 27   Shadows cast in December by a 300-foot building (Boston, Lat. 42°N) (Source: Atkinson, W., in Buttin and Berlin, 1980)

Fig. 28   Environmental impacts of transport (Source: Barton and Bruder, 1995)

or thousands of occupants. The imposition of a high-intensity building such as a skyscraper onto a location should be coordinated (where possible) with other buildings to manage the volume of traffic that will be generated and the impact on the roads and mass-transit systems, as well as the energy consequences. Clearly, the layout of the access routes and roads within the site are as much areas of concern as the energy implications of the distribution of goods from our built system and of bringing materials to the site during construction. The energy used per mile per ton of freight (by road in the United Kingdom) averages 0.0056 GJ (Woolley et al., 1997, p. 56).

Global pollution	Local pollution	Resource use	Aesthetic impact	Physical impact
Emission of greenhouse gases CO_2, NO_2 and (indirectly) O_2	Health and fertility effects of carbon monoxide	Increasing use of the limited resources of oil	Visual impact of roads, etc. on town and country	Intimidation of pedestrians and cyclists
Contribution to acid rain SO_2, NO_2	Child mental development effects of lead	Use of mental and other non-renewables in manufacturing	Visual impact of parked and moving vehicles	Accidents
	Health effects of SO_2, black smoke, volatile organic compounds (VOCs) and low-level ozone	Use of scarce land resources for roads and parking	Noise, vibration and fumes	Creation of barriers to the movement pedestrians

In setting out the site layout plan, we should encourage people to walk instead of using cars. Lazy walking distance standards are about 150 metres with a maximum walking distance of, say, 300 metres. If walking as a mode of transport is to be encouraged, then the layout must provide routes that are wide enough to cope with the population density. If covered verandas are preferred (in hot humid climates), then these should be provided. Buses, trams or trains are more energy efficient than cars. For instance, trains run on electricity and therefore help to reduce the pollution from petrol exhaust. It has been estimated that 200 trees are required to absorb the carbon dioxide from 1 car (in Zeiher, 1996, p. 28). Therefore, the designer must ensure that the energy impact of the design takes into account car ownership figures: 0.3 cars/person in the United Kingdom, 0.56 cars/person in the United States and 0.0012 cars/person in India (Zeiher, 1996, p. 70).

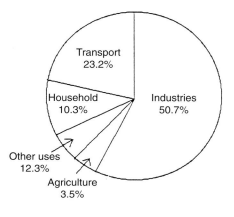

Fig. 29   National energy consumption by sector (Source: Raju, M.K., 1996)

Transport uses over 20 % of the energy consumption in a nation. Therefore the greater the intensification of development, the greater the savings in energy in transportation.

To review, in site planning we should maximise the use of public transport (and create bicycle and footpath systems), avoid further decentralised development (such as new suburban and greenfield developments) and at the same time increase densities and intensities around existing suburban transport routes. The widely dispersed, single-land use development has the effect of separating the demand for goods and services from the supply. To connect the two, people travel increasingly by car, and goods are often transported great distances by freight. An inevitable conclusion is that high-density, mixed-use intensive developments such as skyscrapers help create more efficient, low-energy urban environments.

Landscaping Techniques for Micro-Climate Improvement and Control

The way in which a building is sited relative to other buildings or to natural features of the landscape can be a major determinant of its energy efficiency. The strategy varies, of course, depending on the climate zone. In a temperate or cold climate where protection is needed from cold, the building can take advantage of natural shelter such as trees or banks. Windbreaks, in the form of trees or walls, can prevent the area immediately surrounding the building in temperate climates in winter from becoming as cold as it might if unprotected from the wind; this reduces the need for heating inside. In a tropical climate zone, by contrast, it could be important to ensure that the prevailing breeze can blow easily through the building, creating a natural cooling ventilation. In cities in temperate zones, the position of surrounding buildings is important, as this can determine the flow of wind currents and therefore the temperature.

By making the most of naturally occurring climatic regulators, it is possible to reduce the degree of dependence on artificial forms of heating, cooling and ventilation in temperate zones. A study of the site's natural features, wind direction and positioning of other buildings will create a picture of the local climate; this tells the designer how its features might be exploited positively or protected against. Solar energy is an obvious resource upon which the designer can also capitalise. The use of solar power as a source of energy is not confined to countries with long hours of sunshine. Buildings can be designed to capture and store solar energy, and this can make a worthwhile contribution to heating requirements for temperate zones even in very northerly climates. The zoning of areas with skyscrapers can also improve energy efficiency, for example by planning residential developments in areas which can benefit from solar power, while storage buildings or parking areas are located in areas receiving little sun. We should minimise land-use separation by integrating building uses and maintaining development densities.

Landscape design is too often regarded as merely a cosmetic afterthought, when in reality it can make a significant contribution to improving the environmental impact of a skyscraper or an urban area. Site planning should create green spaces in the skycourts of the skyscraper to improve urban micro-climates and to enhance air quality. We should also use dense planting to protect the edges of the city.

Energy-conscious design requires careful analysis of the natural benefits and problems of a site and the incorporation of features such as slopes, trees, hedges and green mounds into the building plans. The increased use of large areas of vegetation in the upper parts of the skyscraper is to be encouraged, not just for their visual appeal and protection properties, but also because of their ability to absorb carbon dioxide.

For example, in temperate zones, the following techniques, where appropriate and feasible, should be embodied in the landscape design of the skyscraper and other large buildings:

● Solar radiation interception

Planting can block or filter the sun's rays into the interiors of the skyscraper. Deciduous plants can provide beneficial summer shade but
will not impede winter sun. The choice of species will influence the
nature of solar interception, i.e. a plant with dense foliage and a tree
in the skycourt with a tall crown provides more dense shade than
light foliage or small-crown trees. 'Filtering' solar radiation can offer
a more subtle control of light and heat inside the skyscraper than, for
example, sunshades and blinds.

● Solar radiation reflection and control

Grass and planted surfaces around the skyscraper and in all its skycourts reflect less solar radiation than smoother, lighter surfaces.
These grassy surfaces can reduce reflection by at least 60 percent.
In this way, plants will provide insulation for the skyscraper and
absorb heat during daytime, releasing it slowly into the evening,
and therefore moderating external and internal temperatures.

   The texture and colour of the ground and the skyscraper's surfaces can reflect or absorb heat. Reflective surfaces should be minimised to avoid wasted energy, glare and discomfort. Dark-coloured,
absorbent surfaces can store heat and release it later during the
day to extend the use of outdoor spaces and reduce heating
requirements in buildings in temperate and cold climate zones.

● Solar access

For temperate zones, full availability of winter sun can be provided
by avoiding shade from the skyscraper, other structures or planting.
Orientation and slopes are important factors influencing winter
solar gain, taking into account minimum spacing between skyscrapers in an intensive urban environment.

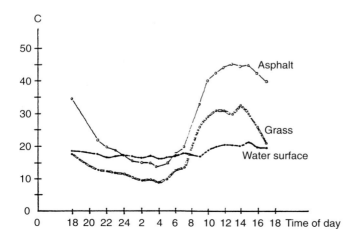

Fig. 30   Surface temperature of materials (Source:
Meiss, Michael (1979), 'The
Climate of Cities', in Ian C.
Laune (ed.), *Nature in Cities*, New York: John Wiley
(1979))

⬤ Temperature control

Temperature is linked directly to solar radiation, wind and precipitation. Shading reduces temperature, but temperature is influenced by plants even if they are not tall enough to provide shade, due to reflection of light, radiation and evapotranspiration.

Planting can trap warmed air beneath a canopy so that night heat loss is reduced and the micro-climate is made more equable. 'Heat sinks' can be achieved with dense foliage, which will contain warmed air to either improve the micro-climate of outdoor spaces or act as an insulating layer next to the skyscraper.

Design of cold air 'drainage' (determining routes of cold air movement) should allow filtering of cold air through permeable barriers or gaps (see Fig. 30).

Landscape can be used to reduce the heat-island effect by providing natural or artificial cover (such as trees, covered walks and planted screens) to the horizontal and vertical surfaces of the skyscraper, and by shading at least 60 percent of impermeable surfaces of the property. Another technique is to provide light colour (high albedo, or reflectance) for 80 percent of the roofing material of a building (e.g. material should have an albedo reflectance of at least 0.5).

⬤ Shelter belts

The designer should aim to filter and therefore reduce wind speed, especially at the ground plane of the skyscraper.

Solid barriers (as against filter barriers), however, create turbulence and eddies. Shelter effect is determined by the height of a shelter belt (a structured combination of trees and shrubs forming a belt of a linear block) and its permeability.

Tree belts and avenues of trees with no shrub understorey can cause wind 'jetting' and wind speed can actually increase by as much as 20 percent. Widths of belts are not as critical as porosity and height. Plant species can be selected to obtain quick establishment, to give the desired year-round permeability and to achieve long-term effect.

The micro-climate near shelter belts improves due to the increase in air and soil temperatures and relative humidity. Orientation of belts should account for varied wind directions. Although there is little scope for planting shelter belts on-site, it is important in temperate zones that the shelter offered by existing woodland blocks the south and east, and belts of trees to the north are maintained by management involving progressive replanting. On-site planting of trees with open canopies will contribute to site shelter.

● Windbreaks

Principles of artificial windbreak design are similar to those of shelter belts. Windbreaks can be located in the skycourts in the upper parts of the skyscraper; 50 percent porosity is best to reduce wind speed over the greatest distance to leeward, and 25 percent porosity is the optimum for lowest wind velocities nearest the windbreak. Windbreaks can be walls or screens, which can be integrated with terrain shape, the skyscraper's built form and the planting system. The porosity of these elements should be designed to either slow down wind without creating turbulence or produce still air pockets. Windbreak devices also could be used temporarily to protect new planting. Walls or special windscreens that are permeable can have a modular or dimensional relationship with the skyscraper's built form. An overall uneven profile is recommended for effective shelter.

● Wind steering

Harmful and uncomfortable winds can be guided by ground shaping, planting, windbreaks or structures in the skyscraper itself. Cold winds can be slowed down, channeled and 'steered' as desired, for example with dense evergreen edges to planting in the skycourts or with ground-shaping at the ground plane. Windbreaks can take the form of fences as sculptural features within layouts, whose purpose is to modify local conditions – especially in atriums, at the facades of the skyscraper or with the location of wind-wall openings.

● Water

Water bodies radiate less solar radiation than the ground surface and therefore act as a moderating influence on temperature changes. The natural flow of cool air from water to land during the day and in the reverse direction at night can be used for ventilation, increasing the comfort of outdoor spaces as well as improving energy conservation.

● Ground shaping

The building levels and associated site works can be organised to minimise the production of surplus spoil. Ground shaping can be particularly effective at channeling winds and breezes at the ground plane of the skyscraper and at the roofscapes. When combined with planting, in improving shelter, spoil will be disposed of economically and constructively by forming part of the shelter pattern and can integrate the skyscraper (including car parking) with the landscape by enclosure or 'sinking' it into the ground. Landscape design should have erosion and storm water control measures (see below).

● Active intervention

Seasonal changes in the micro-climate can be modified to protect areas as required, using a combination of static and mobile landscape elements. For example, the layout of outdoor skycourts could be designed to accommodate screens which can be moved to different positions and alignments depending on the time of year.

● Surface 'roughness' and 'collective shelter'

While there are many complex factors influencing micro-climate, the combination of many varied surfaces and features to form interfaces or 'roughness' at the facades of the skyscraper will contribute to the system of skycourts and atriums in its built form.

● Storm water

Rainwater should be controlled 'at source' (see chapter 4 on outputs). This means that runoff should be detained where it falls and discharged from the area as slowly as possible. This technique will alternate the flood hydrograph. Preferably, when new development takes place, the volume of storm water discharge must not be increased (i.e., aim for 'zero runoff'). When significant volumes of water are detained on the land, flood levels and flood peaks will diminish. Water can be used to charge aquifers and irrigate vegetation, as well as to create ponds, wetlands and water features. Detention basins can also be used to improve water quality by allowing filtration, sedimentation and biological assimilation (Turner, 1998, p. 301).

Whenever possible, rainfall should be infiltrated into the ground where it falls. It should not be ducted into pipes, conduits, ditches or risers. This ensures that the volume of peak discharge is not increased and the interval between the start of a storm and the time of a peak discharge is not reduced. Infiltration enhances soil moisture, regenerates base flow in streams and improves water quality with the restriction of the discharge of pollutants into the drainage network. In the case of the skyscraper, which has a small roof area, the retention of rainwater is mainly at the ground plane, but also from the roof and from all terrace areas and skycourts.

The Urban Ecosystem

The 'systems' of nature are inherently cyclical processes. Nature recycles resources in familiar ways such as food chains. Another example would be the breakdown of decaying plant matter by organisms, which enriches soil and promotes the growth of new plants replacing those that decomposed. The reuse of materials and energy allows the biosphere to maintain its stability and continue to exist over time. As I have stated elsewhere, there is an analogy between the flow of resources in a natural system (an ecosystem) and in a built environment (as a synthetic system) (Yeang, 1972). The use of materials is not a one-way street. This cycling of materials in the natural environment is one of the most salient features that define what an ecosystem is (Bormann and Liken, 1967; Odum, E.P., 1971).

Since the ecosystem will serve as a model for the 'green' built environment, it is worth looking at in some detail. Ecologists suggest that there is a parallel between the dependence of living things on energy flows and an ecosystem's dependence on a cycling of materials and energy (Ovington, 1964; Kormondy, 1969). These principles can be validly applied universally to organisms and ecosystems (Odum, H.T., 1971), and they are foundational to the ecosystem model. This model uses boxes to indicate variously the components of the ecosystem, from the level of decomposers and the materials they break down on up the scale. Energy flows into the ecosystem through photosynthesis and flows out by respiration. In reality, any element or individual within the system can occupy more than one box, which is to say that the roles within an ecosystem are complex and changing. The consumers at the top of the food chain die and become the material for the decomposers to work on; such is the cyclical nature of nature.

The cyclical flow of materials is basic to ecosystems, and yet in most environments the cycle remains incomplete (Sjors, 1955). The further up the hierarchy of ecosystems one goes, the greater is the efficiency – mature ecosystems are more able to retain material for recycling through the ecosystem (approaching a closed loop) than a developing ecosystem. The biosphere as a whole, of course, is a closed system, but its component ecosystems are not. Producers bring energy into an ecosystem; it is eventually dissipated (partly) from the system. The degree to which the circle can be closed indicates how far the ecosystem has gone in its development toward maturity (Odum, E.P., 1969).

In the ambient ecosystem of the skyscraper, which exists in an urban context, one finds a small producer base which is hardly important to the consumers, who are primarily humans. This is an ecological way of saying that plant life found in intensive urban developments is negligible as a food source and has a relatively minor role in the flow of energy through the ecosystem. A city is a dependent system; it draws its energy from other ecosystems external to it, for example producers in rural-agricultural areas and, of course, from the dead ecosystems which power our cities thanks to their decomposition into fossil fuels (Darling and Dasman, 1972; Hughes, 1974).

Urban ecosystems are less stable and predictable than mature natural ones, from which they are poles apart. The existent city, instead of being cyclical in its use of energy and materials, filtering them down to decomposers who can modify them for reuse, operates in a linear, one-way fashion. An enormous flow of imported energy takes the place of a natural cycle. A few natural ecosystems such as rivers provide an analogy to the city because of their linearity and sensitivity to one-way flows which are not recuperated and recycled. The more broken and linear a cycle becomes, the more dependant is the ecosystem to imported energy and materials, because the internal flow no longer provides its needs. The importation of matter and energy may seem 'free' from within the ecosystem (like the flow of fuel oil that once seemed endless), but of course, because of the interrelationship of all ecosystems in the biosphere, there is no such free lunch. The linear, one-way consumption of resources in the manmade environment has damaged and fractured many ecosystems, causing effects such as the exhaustion of soils, drawing down of fossil-fuel resources and extermination of species.

The role of the decomposer strata in the urban ecosystem has also been drastically curtailed. This role is not integrated with the ecosystem and is wholly inadequate. Thus, one ecosystem component can be overwhelmed by the hypertrophy of another or by disruption. In the natural ecosystem, decomposition is carried out by a complex variety of organisms (including fungi and bacteria), all of which act as decomposers (Billings, 1964). But these decomposers actually occupy various trophic levels, and each breaks down dead matter at all the levels at and below it. Without decomposers, the outflow from the system (loss of material that should have been recycled through decomposition) becomes greater, leading to the production of 'wasted' resources trapped in dead animals and plants and with no decomposers to break them down and return

Clearly, the human capacity to consume energy and materials and create waste and nature's capacity to recycle the waste and produce new energy sources and materials are on a collision course.

the much-needed and sometimes rare materials to the system. In existing cities, material that should have cycled within the ecosystem or between systems is simply dumped.

The rapidity of human consumption and the speed of natural regeneration are vastly divergent. (Imagine, for example, human beings waiting for more fossil fuels to be produced.) A key element in green design is not to make the built environment totally dependent on the natural one for the recycling of its outputs, for it is not capable of doing so. It is possible, however, to accelerate natural regeneration rates by human action, and this clearly needs to be done – if it can be accomplished without even further environmental harm. One action that the ecologically minded designer can take is to supply an adequate decomposer layer to his or her built environment. This amounts to imitating in a more faithful way the cyclical pattern of nature, divergence from which has caused many environmental and human problems. Decomposition, production and consumption have to be commensurate. The manmade environment must shoulder some of the responsibility for recovery and not just depend on nature to assimilate and recycle its outputs, for that capacity has obviously been exceeded.

In summary, stage by stage of its life cycle, the skyscraper and other intensive buildings should be examined using the interactions framework to check the impacts on the ecosystems of each activity taken at that stage. In the design of the relationship between the skyscraper and other large buildings and their environments, we can identify three basic strategies: the designer can attempt to control the ecosystem processes of environment in relation to the building, or succumb to them, or seek to cooperate with them. In the last case, the constraints, restraints and inherent opportunities of the ecosystems of the locality and the hinterland of the urban complex have to be examined closely by the designer in the initial stages, following which all design efforts have to be based on the understanding of the ecosystem and directed toward finding compatible combinations and interactions between the natural and built environments.

But in most instances (except for the case of very large plots), the site for the skyscraper is very small – generally under 0.2 hectares (see chapter 1). For such sites, the ecosystem will have been already been significantly devastated and nothing of the site's original ecosystem (its abiotic and biotic components) remains, except perhaps the topsoil (and then, only if there had been no basement on the site), its groundwater conditions, its bedrock and the surrounding air, for example, a zeroculture site. In such cases, the environmental context for building becomes the city block, as well as the entire city and its hinterland. A major exception is in the developing world, where large cities are often laid out on greenfield sites with intact ecosystems.

The environmental factors for ecological design are influenced by the built form and in turn influence the operation of the built form. In the ecological approach, the designer must see the environment in which the designed system is to be located as more than just a spatial zone with physical, climatic and aesthetic features. It exists as part of an ecosystem whose abiotic and biotic components and processes must be taken into account in the design. The layout and location of the built system also depend upon these environmental factors. In addition to these considerations, attitudes to the environment as an infinite source of resources and sink for waste have to be changed; we must acknowledge nature's limitations, particularly in its regenerative capacities.

The external dependencies of a skyscraper (i.e. its environmental dependencies, or L22) consist of the totality of the ecosystems in the biosphere and the earth's resources. The skyscraper and other intensive buildings spatially displace the ecosystems by their presence and deplete the earth's energy and material resources by their creation, operation and final disposal. The use of terrestrial resources for the building further involves extensive modifications to the ecosystems at the points where resources are extracted from the earth or made available. Thus, the designer is responsible for thinking about where the materials for the building are to come from.

Chapter 3 developed the interactions framework which accounts for the activities and processes inside and outside of the skyscraper and the interactions between the two. Here I have stressed the importance for the designer of mastering these impacts; the green approach to design takes into account the effects that all architectural projects and creations have on the earth. Beyond the brute fact of spatial displacement – something is there that wasn't before – there are more subtle and far-reaching effects. The resources and energy that go into the building of the skyscraper and other intensive buildings change the ecosystem's inventory of resources, energy and materials; during its use, the structure's existence and operation may facilitate or cause other human actions (most prominently further development). Hence, a design once executed can in a sense generate other structures, greater population, more resource use and other unintended or unforeseen developments.

Not only does it take energy and materials (in very large amounts) to build a skyscraper; it requires the use of energy and resources just to extract or process these materials. Extraction has its own attendant ecological effects, which though usually taking place off-site (in terms of the skyscraper) are also a part of the skyscraper's total economy of resources and effects, and must be seen by the designer as such. Then, of course, there is the more obvious form of environmental impact that comes to mind when one thinks about manmade systems: a skyscraper discharges pollutants and waste energy in the form of heat into the surrounding environment. Just as it consumes large amounts of material, a skyscraper's outputs are also large, and their effect on ecosystems both local and global can be correspondingly great.

The large building also has effects on other manmade systems, such as built environments and infrastructure. This is especially true in the usual skyscraper context, the densely built inner city. For example, its shadow may affect the ambient temperature and solar energy productivity potential of other buildings. One approach to mitigating this effect is the use of the "solar envelope" concept. The "solar envelope" is defined as the largest hypothetical volume that can be constructed on a lot without overshadowing neighbouring properties during critical energy-receiving hours and seasons. It is therefore both a temporal and spatial calculation. Studies indicate that this concept in urban contexts is effective for plot ratios up to 1:6 (Steele, op. cit.).

The primary focus of ecological site analysis prior to building our designed system is to ensure that the land is used efficiently and that the building fits appropriately into the service infrastructure. This section, hopefully, has served to remind the designer that the ecology of the project site must be taken into account in the design of the skyscraper. Where the project site is already ecologically devastated, then design should seek to repair the site's earlier functioning ecosystem and to enable recolonisation to take place in a manner integrated with the surrounding ecosystems. Site ecological factors have been listed (see above) by category for the designer's convenience, and can be placed under broad headings such as species and populations (plants and animals), habitats and communities and ecosystem processes.

We have also noted that the natural resources available in a given locality include its climate, which can be harnessed for passive low-energy heating and cooling purposes through bioclimatic design (see above and chapter 7). The elevation of the sun, seasonal and regional range of sunlight, air temperatures, wind force and direction, periods when winds occur, quantities of precipitation, the degree of exposure and aspect of open spaces, the ground's angle of slope and contour, topographical formations, areas of water and vegetation are some of the features that deserve the designer's attention.

Next we will move from discussion of the ecosystem properties of the project site to the large number of the design decisions with 'off-site' impacts. The site is part of an ecosystem, but decisions made there affect other places. Materials are one source of such effects (dealt with in chapter 6); they may be taken from environments where biodiversity is decreasing (such as tropical forests or unsustainable native forests). All site planning must ensure that there is no loss of species deliberately or by default. Design strategies should seek conservation more than replacement (a more difficult and dubious proposition), and when replacing, to use indigenous rather than non-native or exotic species. The general design strategy is to conserve the highest level of biodiversity possible; we should restore biodiversity whenever it is has been removed and increase the diversity of all degraded sites.

Chapter Six

Design Regarded as Management of Energy and Materials

Building is a natural function. When building activities take place in nature, the organism (whether it be a bird, ant, rodent or other animal) takes materials from a variety of nearby sources and re-concentrates them in one specific location (the 'project site'). These materials are assembled into an enclosure for the organism's activities and to protect it from the climatic elements or from other, hostile organisms. It is an inescapable fact that all building activities involve the use, redistribution and concentration of some component of the earth's energy and material resources into specific areas, with the effect of changing the ecology of that part of the biosphere as well as adding to the composition of the local ecosystem. As we have mentioned earlier, the continued existence and maintenance of the built environment involves dependence on the earth's resources and environment, which must supply it with certain inputs (see Fig. 31).

In the case of humans, the distances from which materials are taken and transported are intercontinental and the intensiveness of scale of enclosure (as in the case of the skyscraper) is not only large but unprecedented. The 'inputs' into the built environment (see chapter 3 with reference to L21) include not only construction materials but also the energy derived from non-renewable sources to effect the transportation of the materials, their assembly and construction on the site as well as the energy required to sustain the internal environmental conditions through the operational systems (see chapter 7 with reference to L11) to their eventual recycling or reuse or ecological reintegration. However, the consequences of the above are not just on the project site's ecosystem, but on the global environment as a whole. Both outputs and inevitable wastes are emitted (see Fig. 32 or L12 in the partitioned matrix in chapter 3).

For these reasons, the designer might regard the creation of a building as a form of energy and materials management or as prudent resource management. For example, the supply of electricity involves the conversion of fuel into energy, and this conversion process depletes non-renewable resources; in addition, it can have a lasting negative effect on the environment through the related emissions caused over the entire life cycle of the building. As we

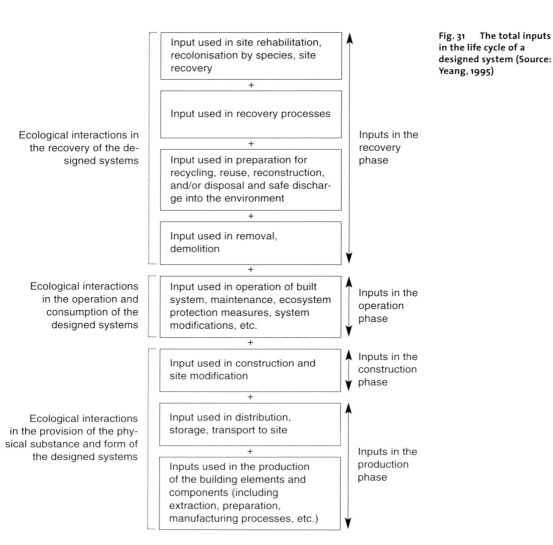

Fig. 31   The total inputs in the life cycle of a designed system (Source: Yeang, 1995)

The diagram contents:

Ecological interactions in the recovery of the designed systems

Inputs in the recovery phase
- Input used in site rehabilitation, recolonisation by species, site recovery
+
- Input used in recovery processes
+
- Input used in preparation for recycling, reuse, reconstruction, and/or disposal and safe discharge into the environment
+
- Input used in removal, demolition
+

Ecological interactions in the operation and consumption of the designed systems

Inputs in the operation phase
- Input used in operation of built system, maintenance, ecosystem protection measures, system modifications, etc.
+

Inputs in the construction phase
- Input used in construction and site modification
+

Ecological interactions in the provision of the physical substance and form of the designed systems

Inputs in the production phase
- Input used in distribution, storage, transport to site
+
- Inputs used in the production of the building elements and components (including extraction, preparation, manufacturing processes, etc.)

have noted, green design takes account of the entire process and operation of the built system, including its final dismantling; in effect, the ecological designer designs the life and death of his creation. In ecological design, he or she is ethically responsible for the building over its life cycle and for what happens to the building and its materials at the end of their useful life. Therefore, the designer must not only understand all environmental implications, and all building inputs and outputs, but also must seek to recycle all outputs as inputs to other processes. The designer must be concerned with how our skyscraper and other intensive building types and all its component parts can be disassembled in ways that will allow maximum levels of recycling (see Fig. 33).

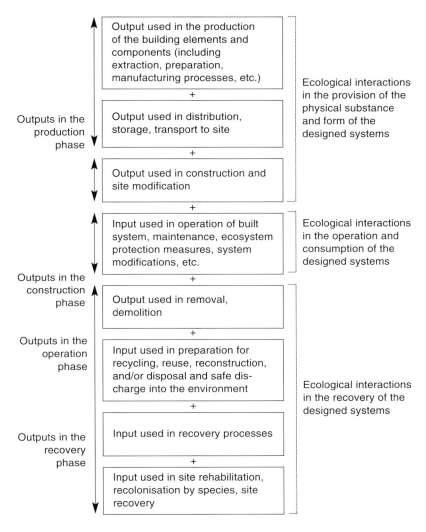

Fig. 32   The total outputs in the life cycle of a designed system (Source: Yeang, 1995)

Outputs in the production phase

Outputs in the construction phase

Outputs in the operation phase

Outputs in the recovery phase

Output used in the production of the building elements and components (including extraction, preparation, manufacturing processes, etc.)

+

Output used in distribution, storage, transport to site

+

Output used in construction and site modification

+

Input used in operation of built system, maintenance, ecosystem protection measures, system modifications, etc.

+

Output used in removal, demolition

+

Input used in preparation for recycling, reuse, reconstruction, and/or disposal and safe discharge into the environment

+

Input used in recovery processes

+

Input used in site rehabilitation, recolonisation by species, site recovery

Ecological interactions in the provision of the physical substance and form of the designed systems

Ecological interactions in the operation and consumption of the designed systems

Ecological interactions in the recovery of the designed systems

Design Responsibility for the Long-Term Fate of the Built System

The designer's responsibility extends even beyond the point at which the building is handed over to the client or user. From an environmental perspective, the designer is ethically responsible for the disposition of the materials in a building from 'source to sink'. Working within the parameters of a rigorous ecological approach, the designer determines not only how and at what environmental cost (in terms of demands for materials and impact on the planet) a designed system can be built, but he or she also analyses how and at what cost to the environment the built work will be used, managed and ultimately disposed of (see Fig. 34). In other design fields

129

this approach, embodying both initial design and subsequent use and disposal, is called 'DFD' (Design for Disassembly) (in Papanek, 1995).

As contemporary law and other determining professional institutions have conceived of the design process, designers are not responsible for the later disposal of the components of their works.

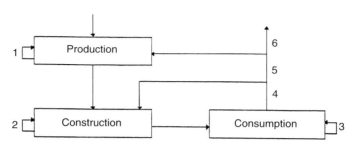

Fig. 33   Strategy for the cyclical pattern of use of building materials (Source: Yeang, 1995)

Where
1. Recovery within production processes of the building materials
2. Recovery of construction residuals from construction processes
3. Recovery from operations of the built system
4. Recovery of materials from operations into construction processes
5. Recovery of materials from operations into production processes
6. Redirected use of materials elsewhere

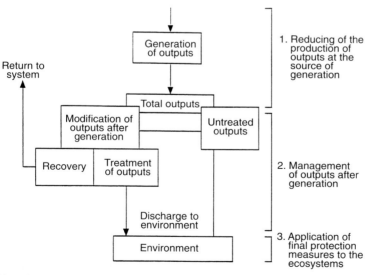

Fig. 34   Generation of outputs (Source: Yeang, 1995)

Map of the possible routes taken by the outputs from the built environment. This is a generic model for tracing the flow of outputs from the built environment; it could be considered as a problem definition tool. No equivalent status is accorded to each of the measures listed. The selection of the appropriate form of management for each individual output from a system will thus depend on the built system, the form of outputs discharged, the operating conditions, the form of the inputs, the state of the environment, and the interaction of all these factors.

Fig. 35 Generalised flow of materials, use, reuse and recovery (Source: Ghazali, Z. M., Kassim, M.A., 1994, p. 186)

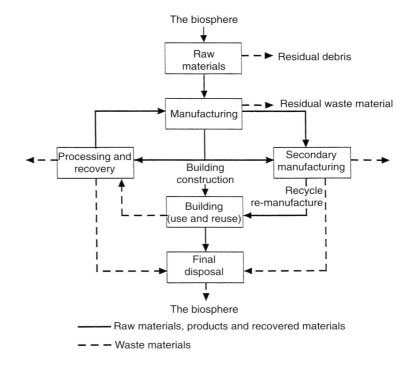

The long-term ecological impact of their efforts is not traditionally the responsibility of designers. One possible solution to environmental problems would be to make manufacturers or suppliers of building materials and components responsible for these long-term consequences. Under such a system, the designer would assemble the materials and components in a building with the understanding that these would eventually be 're-manufactured' or returned to the supplier or manufacturer.

Customarily, when a building reaches the end of its usefulness, it is demolished and the constituent materials are discarded or salvaged for reuse elsewhere. In the present economic system, materials flow one-way from their points of origin as natural resources, through their transformation and assembly into goods, to their sale to the consumer. At this point, the environmental impact of building materials and other components is forgotten as they are assembled into a larger work. The consumer uses the building but it is not 'consumed', strictly speaking: he or she simply discards it after use. Many products that are ordinarily considered to be 'consumed' actually only render temporary service in the built environment. Even when a building is demolished, its fabric remains as output to the environment in the form of discarded materials.

Building type	Embodied energy Delivered GJ/m^2	Embodied energy Primary GJ/m^2	Embodied CO_2 kg CO_2/m^2
Office	5–10	10–18	500–1000
House	4.5– 8	9–13	800–1200
Flat	5–10	10–18	500–1000
Industrial	4– 7	7–12	400– 70
Road	1– 5	2–10	130– 650

Fig. 36 Overall embodied energy and embodied CO2 for various building types (Source: Yeang)

If this one-way flow of materials persists as accepted practice, the rampant discarding of waste building fabric will result in its accumulation in the ecosystem and this mass will subsequently strain the environment's carrying capacity. Even those elements that do not compound disposal problems during their periods of use will still exacerbate environmental problems when discarded later. From the perspective of the life cycles of materials, works of architecture and other built structures must be seen as potential waste. The eventual reuse of materials needs to concern designers from the outset (see Fig. 35).

Designers obviously do not like to dwell upon the idea that their buildings will one day be demolished, but it happens anyway sooner or later. In the ecological approach, the designer must assess the potential for the building's components to be recycled. While potential is not the same as actuality, such considerations assist designers in becoming aware of the need for their own participation in processes of recovery and reuse. These are the end objectives of our role as designers in the management of materials and energy in the skyscraper and our large building type. From the point of view of the ecologist, a building is simply a transient phase in the flow of materials and energy in the biosphere, managed and assembled by people for a brief period of use (a period usually economically defined).

The construction of a designed system such as a skyscraper requires significantly large inputs of energy (see Fig. 36) and materials from the environment. Theoretically, these constitute the external-to-internal interdependencies of the built environment (i.e. L21 in the partitioned matrix), and include inputs needed for its maintenance and for the disposal of its outputs. These include not only the energy and materials used to synthesise its physical substance and form (see Fig. 37), but also those that are used to maintain it in all the phases of its entire life cycle. During its operation, outputs are emitted and other impacts are inflicted upon the ecosystem. If we build with salvaged materials, we conserve more than half of the energy investment. In the ecological approach, the designer must map and quantify all external-to-internal exchanges and any envir-

Fig. 37 Process energy requirements (PER) for common building materials (Source: Lawson, 1996)

onmental impacts resulting therefrom. From a holistic point of view, the use of inputs into the built environment is related to the discharge of outputs, the set of operations in the built environment and the limitations of the earth's ecosystems and resources.

The interdependencies between the built and natural environments include the obvious spatial displacements of portions of the ecosystems caused by building; the amount of energy and matter used in and by the designed system; its emissions of energy and output of matter; and the consequences of human activity within the built environment. In the largest sense, the ecological consequences of the built environment must be considered to be not only those immediately related to the construction of buildings and other artefacts, but must also be understood to include all of the environmental interactions that derive from the use of built artefacts, their later disposal when no longer useful and their eventual recovery.

Building inputs	
Material	Embodied energy MJ/kg
Kiln-dried sawn softwood	3.4
Kiln-dried sawn hardwood	2.0
Air-dried sawn hardwood	0.5
Hardboard	24.1
Particle board	8.0
Medium density fibreboard (MDF)	11.3
Plywood	10.4
Glued-laminated timber	11.0
Laminated veneer lumber	11.0
Plastics, general	90.0
PVC	80.0
Synthetic rubber	110.0
Acrylic paint	61.5
Stabilised earth	0.7
Imported dimension granite	13.9
Local dimension granite	5.9
Clay bricks	2.5
Cement	5.6
Gypsum plaster	2.9
Plasterboard	4.4
Fibre cement	7.6
In-situ concrete	1.7
Precast steam-cured concrete	2.0
Precast tilt-up concrete	1.9
Concrete blocks	1.4
Autoclaved aerated concrete (AAC)	3.6
Glass	12.7
Mild steel	34.0
Galvanised mild steel	38.0
Aluminium	170.0
Copper	100.0
Zinc	51.0

As discussed earlier, nearly all existing design has been based on the mistaken assumption that the earth's natural resources (i.e. raw materials, fossil fuels, land, and other materials) are infinite and that the planet functions as a limitless sink for the disposal of any wastes that humans generate. The relationship that commonly exists between the built and natural environments is open-ended and linear. This existing pattern might be described as a 'once-through' system in which resources are used at one end, and wastes are expelled at the other. The designer has usually considered resources mainly at the point of use. It might be suggested that just as the application of technologies may have inadvertently exploited

the biosphere, so technologies, if employed with a fuller understanding of the ecological systems, could result in designs through
which humanity could live in a better balance with nature.

In order to conceptualise the consequences of 'energy and materials management', we can think of the process in terms of specific
patterns of energy use and the flow of materials from their origins
in the environment, through their incorporation into buildings and
lastly to their eventual ends in environmental sinks. By considering
the process thus, we are able to analyse all constructed objects in
terms of the patterns of use they represent. We can anticipate in
design the extent of demands and impacts that each individual
pattern will exert on the earth's ecosystems and we can predict its
likely future use of resources. The actions and activities that will
take place within a designed system over the course of its life cycle
will determine the flow of inputs into the built environment and
the expelling of outputs into the environment beyond, and will
have a specific impact on the earth's ecosystems as well as on its
supply of resources. For example, about 50 percent of all the energy
consumption (i.e. inputs or L21 in chapter 3) in a country (e.g., in the
United Kingdom; after Edwards, 1996) and a similar proportion of
carbon dioxide emissions (i.e. outputs) is associated with buildings.
The majority (about 60 percent) is used in residential stock, the
remaining 40 percent being divided between offices (7 percent),
warehouses (5 percent), hospitals (4 percent), retail (5 percent),
educational buildings (7 percent), sports facilities (4 percent), and
hotels and other structures (8 percent). These figures indicate that
the major building type that our energy conservation efforts are to
be directed at is primarily the residential building type (60 percent), followed by the office building type (7 percent). The strategies
for the selection of materials in our green skyscraper or large building design are discussed below.

Once the schematic design for the building has been generally
finalised, its physical components should be quantified and
assessed for their anticipated impacts on the environment and for
whether the design meets current standards for green design. In
most conventional building situations, a set of 'Bills of Quantities' is
usually prepared for the entire building (usually for bidding purposes). This quantification can easily be converted into weight
equivalents (e.g. tonnage) of material, or into volumetric quanti-
fications which can then be subsequently converted into energy-
embodied equivalents and ecological-impact-embodied equivalents
for comparative analytical purposes. This can similarly be done for
the operational systems (M&E equipment) in the building.

Extraction	Brick-making	Transport	Use	Reuse/disposal	Implications
Agriculture land loss at clay pit	Air pollution and CO_2 produced at firing	Energy use in transport of bricks	Energy use at construction site	Landfill site needed for disposal	Specify local sources of brick
Ecological impact of extraction	Run-off into water courses	CO_2 produced in transport	Noise at construction site	Brick recycled if possible	Specify mortar mix which allows reuse of bricks
Non-renewable energy use	Non-renewable energy use	Pollution caused by transport of bricks (nitrogen oxide, etc.)	CO_2 produced at construction site	Brick crushed for aggregate	Use brickmakers who follow good environmental practice (see Salveen case study)
Landfill sites created for waste disposal	Adverse visual impact of brickworks	Community disturbance in transport of bricks			Exploit wildlife/ amenity potential of clay pits
Water habitats created for wildlife and amenity					

Fig. 38   Life-cycle impacts of a brick (Source: Edwards, 1996)

Strategies for the Selection of Materials

The designer of the skyscraper and other intensive building types has to select materials for the building (i.e. L21). Besides the usual architectural criteria (aesthetics, costs, etc.), the ecological criteria for selection should be as follows:

● The potential of the material for reuse and for recycling (as a consequence of replacement due to wear and tear or at the end of the building's useful life).

● The embodied ecological impact of the material (the impact on the environment as a consequence of the production and delivery of the material to the construction site).

● The embodied energy (the energy cost of producing and delivering the material or component to the construction site) in the material (see Fig. 38).

● The toxicity of the material on humans and the ecosystems.

The point here is that the selection of materials is not as straightforward as setting out a priority list of preferred items using a system of embodied energy 'weighting' of different material characteristics (see Fig. 39). Choice also depends on design (e.g. whether designed for recovery). The basis for the materials selection is explained as follows:

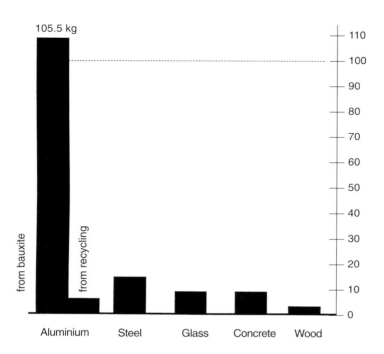

105.5 kg

from bauxite

from recycling

Aluminium   Steel   Glass   Concrete   Wood

Fig. 39   Energy cost of material (Source: Von Weizsacker et al., 1997)

Potential for Reuse and Recycling

Materials must be selected based on their potential for reuse and recycling. In regarding design as the management of the inputs and the outputs of the building, our concern has to begin with the objectives (being the end consequence of the management activity, hence referred to as MBO or 'management by objectives'). The objectives are simply to reduce the impact on the natural environment and positively to renew, restore and enhance the natural environment. This is best achieved by retaining the materials and components of the built environment by reuse. Materials with potential for reuse must be given priority over materials with potential for recycling, as reuse uses less energy and effort, but both avoid discharge into the environment.

A material's potential for reuse and recycling might be given greater priority over the level of the 'embodied energy' value. There is a wrong belief that priority should be placed on materials and components that have low embodied energy values (see Fig. 36, and Demkin, 1996). On the contrary, the potential for a material for primary or secondary reuse is more important, for every time that a material is reused, its original embodied energy is reduced (i.e. less about 50 percent of its original embodied energy, depending on the energy cost of reuse). This means that the more times a material or component is reused, the more its 'embodied energy cost' goes down in value, thereby enhancing its suitability for first use.

As an initial strategy, materials from non-renewable resources should be made as reuseable or as recyclable as possible to conserve the material. At the same time, in materials selection, priority should be given to materials that have been used previously (i.e. 'waste' from an earlier structure) or that have been recycled. This immediately lowers the overall embodied energy figures in the building's mass.

Embodied Ecological Impact

'Embodied ecological impact' reflects the impact of the production of the material or component (both globally and locally) at the source of manufacture, as well as the chain of activities leading up to its delivery to the construction site. An indicator commonly used (Howard et al., 1995) is the carbon dioxide emissions measured in kilograms. But carbon dioxide emissions are not the only ecological impact in the production of a material; others include land devastation, pollution of waterways and energy cost of transportation.

For a more rigorous ecological design approach, considerations should include further considerations:

● The record of suppliers of raw materials and components should be examined to ensure that their manufacturing processes are not unnecessarily polluting. As a minimum requirement, suppliers should be able to demonstrate that they have not contravened local legislation and acceptable ecosystem impacts concerning emissions.

● The processes by which specific ingredients are manufactured should be questioned for the impact on the environment (pollution in manufacture).

● Products (varying from paints to toothpaste) can be produced in different ways with varying degrees of polluting effects. For example, paper produced by mills that use chlorine bleach should be avoided in favour of unbleached paper, or paper bleached using hydrogen peroxide.

- The impact of the materials on the environment after use should also be considered. We need to check whether they are fully biodegradable and whether there will be a contamination problem if the products end up in landfill sites.

- Saving water is as important in some regions as saving energy. Designers can aim to design products and parts that use far less water.

- Materials may be natural, synthetic recycled or virgin, renewable or non-renewable. There are no clear-cut answers about which is least environmentally damaging. The belief that naturally occurring materials are to be preferred over manmade materials can be erroneous as these materials (e.g. tropical hardwoods) may be in very short supply or their production may result in environmental impairment or loss of biodiversity.

- Materials that occur near to their point of use have the advantage that they require less energy to transport. The use of locally occurring construction materials is argued by some as engendering a more critical, regionalist design.

- The extraction process of some raw materials, such as aluminium and gold, can cause severe damage to local habitats. While the designer cannot take responsibility for what happens at the very start of the supply chain, this is one area where information should be requested wherever possible from raw materials suppliers or intermediaries.

Material selection decisions
should reflect the desire to support those materials whose mining and extraction activities cause the least damage.

In aggregate, the embodied ecological impact of a material can be evaluated at three levels: global sustainability, natural resource management and local environmental quality at source of production.

Embodied Energy Impact

Embodied energy in a material or assembled component is the energy expended (from non-renewable energy sources) in the extraction and processing of raw materials, manufacture, transport and construction. Some hold that the embodied energy of a product is calculable with a high error margin (Daniels, 1995). The term is regarded by some as not adequately defined.

It has been found that a major cost of the embodied energy in a material is in transportation to the construction site. The contention is that local products are preferred, as these will travel shorter distances than those from elsewhere; however, this has to be evaluated against the reuse potential of the material or component. A local or regional product that has limited reuse potential may have a lower overall embodied energy value than an imported material that can be reused many times over. Furthermore, the local material may be in limited supply as a resource.

Much of the literature on energy 'contained' within the building materials relates to mass rather than to use, which can be misleading. For example, it may be incorrect to say that glass is a preferred material to brick by virtue of its lower mass, and that therefore an

Fig. 40 Example of embodied energy and CO_2 analysis for an office building (Source: Yeang)

Element	Measured (m^2)	Embodied energy (GJ prim)	Embodied energy (GJ del)	Embodied CO_2 (kg CO_2)	Life cycle embodied energy (GJ prim)	Life cycle embodied energy (GJ del)	Life cycle embodied energy CO_2 (kg CO_2)	Primary % initial	Primary % life cycle
Substructure	13527	155450	98778	16795429	155405	98778	16795429	28%	13%
Structural frame	24481	226882	126659	23593701	242201	139343	24884596	41%	20%
External walling and finishes	20377	40754	20377	3056550	22262	61131	9169650	7%	10%
Roof coverings, etc.	6414	11452	6449	1023177	23481	13352	1922352	2%	2%
Internal walling	944	415	363	35896	2492	2176	215374	0%	0%
Internal finishes	N/A	49587	28623	366561	285424	162046	20052644	9%	23%
Joinery finishes, etc.	49157	0	0	0	0	0	0	0%	0%
Services	49157	62291	46227	3868386	404632	291795	23839758	11%	33%
GJ/m^2	49157	546797 11.1	327475 6.7	42138700 1061	1235897 25.1	768620 15.6	96879803 1971	100%	100%

TRI Chemical	Potential health effects[a]	Sample products
Acetaldehyde	A, C, E[b]	Adhesives
Acetone	Ch, E	Varnishes, lacquers, paint thinners, furniture strippers, adhesives, nail polish remover, art supplies, metal polish
Acrylonitrile	A, C, D, E, R	Fabric, apparel
Ammonia	A, Ch, E	All-purpose cleaners, glass cleaners, hair dyes
1,3-Butadiene	C, Ch, D, R	Carpets
Carbaryl	A, Ch, E, N, R	Pesticides, pet flea and tick treatment
Cadmium	A, C, Ch, D, E, R	Nickel/cadmium batteries
Chlorine	A, Ch, E	Chlorine bleach, disinfectants
Chlorothalonil	C, E	Lawn chemicals
Cresol	A, Ch, E	Art supplies, disinfectants
2,4-D	A, C, D, E, R	Lawn chemicals
Dibutyl phthalate	Ch, D, E, R	Paint, adhesives
1,2-Dichlorobenzene	E	Carpets
Di(2-ethylhexyl) phthalate (DEHP)	C, Ch, D, E, M, R	Plicate plastics, fabrics, apparel, hair spray
Diethyl phthalate	A, E	Paint, adhesives
2-Ethoxyethanol	Ch, D, R	Polyurethane, wood finish
Ethylbenzene	Ch, D, E, R	Carpets, paint
Ethylene glycol	Ch	Deodorants/antiperspirants, paints
Formaldehyde	C, Ch, E, M, R	Plywood, particleboard, clothing, adhesives, upholstery, fabric, fingernail polish
Hydrochloric acid	A, Ch	Toilet cleaners
Lead	D, N, R	Batteries, stain/varnish/sealant, hair dyes
Mercury	Ch, E, N, R	Batteries
Methanol	N	Paint thinner, strippers, adhesives
Methylene chloride (dichloromethane)	C	Spray paint, rust paints, paint strippers, adhesives, adhesive removers, pesticides
Methyl ethyl ketone	Ch, D, N, R	Paint thinner, adhesives, cleaners, waxes
Methyl isobutyl ketone	Ch, N	Paint thinner, pesticides
Naphthalene	E	Mothballs, adhesives, pesticides
Nickel	C, Ch, D, R	Batteries
n-Butyl alcohol (n-butanol)	Ch	Paint strippers, perfume, aftershave lotion
Paradichlorobenzene (p-dichlorobenzene)	C, Ch, E	Mothballs/crystals, certain air fresheners, toilet deodorizers
Pentachlorophenol	A, D, E, R	Varnish/stain/sealant
Perchloroethylene (tetrachloroethylene)	C, Ch, D, E, R	Dry cleaning, rug/upholstery cleaners, spot removers
Phenol	A, D, E	Art supplies, adhesives
Phosphoric acid	None listed	Metal polish
Styrene	C, Ch, E, M	Carpets, building materials
Sulphurcic acid	A, Ch, E	Batteries
Toluene	D, E, M, R	Paint, nail polish, furniture strippers, adhesives, art supplies, carpets, paint strippers and thinners
1,1,1-Trichloroethane (methyl chloroform)	D, E, R	Carpets, dry cleaning, spot removers, fabrics, typewriter correction fluid, adhesives/glues
Vinyl chloride	C, Ch, D, M, R	A variety of household plastics, including furnishings and apparel
Xylenes	Ch, D, E, R	Paints, adhesives, pesticides, art supplies, furniture strippers
Zinc	E	Batteries

(Source: INFORM, Inc., 'Tackling Toxics in Everyday Products: A Directory of Organizations' (New York, 1992))
Note:
a. Health effects data are from the U.S. Environmental Protection Agency's (EPA's) toxicity matrix for right-to-know chemicals. The assembled information is presumed to be incomplete; further testing would be likely to produce additional entries on the matrix. For information on chemical synonyms (omitted here) get the EPA's "Common Synonyms" publication.
b. Key to abbreviations: A – acute toxin, C – carcinogen, Ch – chronic toxin, D – developmental effects, E – environmental toxin, M – mutagen, N – neurotoxin, R – reproductive effects.

Fig. 41   Toxic chemicals in common building materials (Source: INFORM Iwe., 1992)

external wall of glass is preferable to brick. Brick, while having a higher mass, also has better insulation properties than glass, thereby providing energy savings in the operational life of the building. On the other hand, glass may be easier to demount and reuse than brick.

Pre-assembly of materials has also to be factored in. For example, timber-frame construction can be up to 20 percent less energy intensive than traditional timber construction. In a two-bedroom dwelling, the embodied energy is equivalent to between two and five years of energy consumed in its operational phase (by heating, light and power). Thus while embodied energy is one factor in ecological design affecting its first costs, the operational energy costs by far exceed first costs in terms of energy use, but may actually be less when comparing embodied ecological impacts. Some studies in Europe indicate that the production energy (primary energy at approx. 2.0 mWh/m²) contained in a building is equal to approximately 20 percent of the operating energy required over its entire lifetime (Herzog, ed., 1996, p. 35).

Studies have shown that the level of embodied energy in a building is related to its mass. Generally, the lower the building's mass, the lower would be the total value of its overall embodied energy in its materials and equipment. A low-energy-embodied office building should not be more than 10 GJ (Gigajoules)/m² of primary delivered embodied energy (Howard, 1997) (see Fig. 36).

Reinforced concrete-frame construction has almost the same amount of embodied energy as steel, but it is less recyclable at the end of its useful life than steel. Structural steel can generally be recycled and reused virtually in its original use, whereas concrete can be reused mostly in a downgraded form (e.g., as rubble) and, with limitations, can be recycled again for structural purposes.

As embodied energy constitutes only about 35 percent of the energy used in the lifespan of a building (i.e. L21 in the partitioned matrix in chapter 3), the greater amount is used during its operational mode (up to 65 percent, i.e. L11 in the partitioned matrix in chapter 3). This means that we should place greater importance on passive design so that during its operational stage the building will make the most of ambient energy, e.g., natural daylight, natural ventilation, etc., to minimise the remaining energy demands over its lifetime (considering a 60-year building lifespan).

The embodied energy/carbon dioxide calculations are a key indicator of environmentally conscious building design. However, when comparing embodied energy values of alternative construction types, it is not the embodied energy value per unit mass or unit volume of the material that is important. Rather, it is the building that has to be compared, as alternative materials used for similar functions will have different properties. This comparative component is termed the 'functional unit'. For example, although steel has a higher embodied energy per unit mass than reinforced concrete, it is considerably stiffer. The result is that much less steel is required to perform the same structural function than reinforced concrete.

Therefore, when making such calculations, what should be compared is the embodied energy of the 'functional unit'. In the example above, this might be effected by comparing a steel beam with a reinforced concrete beam, which performs the same function. But when comparing frame alternatives in office buildings, the functional unit should be the gross square metre area of the completed building, as many substantive factors result from adopting either a steel or concrete frame, such as the type of foundations and the structure's fire resistance.

Research carried out at the Steel Construction Institute (SCI) (in the UK), which has compared steel and reinforced concrete frame construction for office buildings, has shown that there are no significant embodied energy/carbon dioxide differences between the alternative construction types. Thus, there are no embodied energy/carbon dioxide penalties for designers selecting either material (in *Building Design*, 1998). However, a designer who selects high-energy-embodied materials must make sure to pay particular attention to the recycling or reuse after the end of the useful life of the building.

To facilitate reuse and recycling, consideration also must be given to the method in which the material is fixed to the building structure. Mechanical forms of fixing facilitate reuse, whereas chemical and altered forms of fixing inhibit it.

● Toxicity

Each material, product, and component to be used in the skyscraper or our intensive building type needs to be carefully studied for its technical contents, manufacturing history and performance record.

The intention is to minimise the toxic content of the material and the effect on humans, who after all are one of the key species in the ecosystem of the building (see Fig. 42) and on the ecosystems. For example, office furnishings should be tested for the presence of formaldehyde, and paints with minimal volatile organic compounds (VOCs) should be applied to walls. All carpets, paints, wall

Type	Properties	Use
Permanent	For products for which there will be no secondary use. Applications in medicine and related fields for products in direct contact with organic parts, e.g. parts of an implanted hip-joint, shell of heart pacemaker, artificial veins, blood-storage bags. Materials characteristics and lasting quality performance of primary importance, e.g. nylon 66. Quantity used negligible.	C
Reuseable	Product can be used over and over again unchanged, e.g. plastic bucket. Complex tools or appliances can be repaired, upgraded in whole or in part, for resale. Enormous numbers of items involved. Wood, tin, enamel, glass, ceramics ecologically and aesthetically preferable.	A
Recyclable	Thermoplastics and elastomers melt at a specific high temperature like glass and are easy to recycle. Thermosetting polymers do not liquefy and are very difficult to recycle; research is continuing into better methods.	B
Co-recyclable	Compatible materials can be recycled together to form a useful new material.	B
Biodisintegratable	Attempts have been made to embed a biodegradable trait into synthetic polymers so that they turn into mulch. These compounds perform badly in landfills through lack of moisture, slightly better when composed. Radical improvements have produced plastics, now commercially available, that degrade 100% less than 2 months after being discarded. Research continues into further control of the start of degradation.	B
Biodegradable	100% biodegradable rather than biodisintegratable. PHA (polyhydroxyalkanoate), a member of the polyester family, is 'manufactured' directly by micro-organisms. Scores of bacteria that produce this organic polymer have been found, including PHBs (polyhydroxybutyrates), one of the first to be commercially available. PHA plastics can be moulded, melted and shaped like petroleum-based plastics, and have the same flexibility and strength. The same production methods can be used, e.g. melt-casting, injection-moulding, blow-moulding, spinning and extrusion.	A
Bioregenerative	Polycaprolactone film completely biodegrades within 3 months, leaving no residues. Research into paper products laminated with layers of com-based cellulose materials prove they can resist water for 6 to 8 hours and could serve as containers for drinks and fast-food items.	A
Bioenhancing	Carry additives to stimulate plant growth, prevent erosion in arid climates (artificial burrs), carry plant seeds and seedlings embedded in growth stimulants.	A

Fig. 42 Categories of plastics (Source: Papanek, 1995)

coverings, floor finishes, and furniture systems are to be carefully reviewed to reduce the presence of toxic elements such as formaldehyde, VOCs and other harmful chemicals commonly found in construction materials and which affect indoor air quality.

The variety of chemicals found is usually complex, and as a result, difficult tradeoffs need to be made. The criteria can be prioritised to help in making decisions. For instance, indoor air quality can be given first priority because it affects the building users more directly and because information on chemical composition is more readily available than data on manufacturing processes or disposal. Where particularly dangerous tradeoffs in the upstream and downstream effects need to be made, there may be exceptions (Audubon Society Hq., in Zeiher, op. cit., p. 183). For example, polyvinyl chloride (PVC) plastic emits highly toxic chemicals when incinerated, and thus its use is to be avoided wherever possible.

Establishing a material's toxic chemical content by knowing the quantity of chemicals, the rate at which they emit gas which can be transmitted to building occupants and other hazardous potential is not an exact science. For example, the review of the United States Manufacturer's Material Safety Data Sheets (MSDS) is a first step in determining the presence of harmful chemicals in products. MSDS sheets provided by the manufacturer upon request list the names and amounts of major chemicals found in a product. The information next can be checked against the International Agency for Research on Cancer list of carcinogens and a number of handbooks on toxicology.

However, the lack of information from manufacturers makes general research by the designer on product types a pivotal requirement in the decision-making process. Although MSDS sheets can be found, a certain expertise is required to interpret the scientific data. For example, widely available environmental research on plastics has helped scientists to rank them as follows (see Fig. 43):

Type	Rank
● Polyethylene, polypropylene	Benign
● PET and Polystyrene	Intermediate
● ABS	Questionable
● Polyvinyl chloride (PVC)	To be avoided

Toxic solvents and alkyds such as benzene, xylene, and toluene are found in many common building products (e.g., in oil-based paints) and can enter the bloodstream and cause respiratory problems, allergic reactions and liver damage. Chemicals used in the production of fabrics, carpets and pressed woods used in furniture systems include formaldehyde, diphenol ether and styrene, all of which are

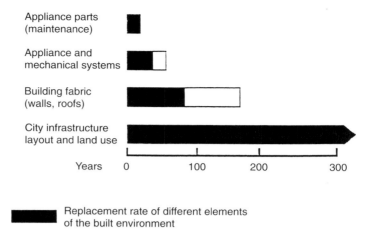

Fig. 43 Replacement rate of different elements of the built environment (Source: Yeang)

Appliance parts (maintenance)

Appliance and mechanical systems

Building fabric (walls, roofs)

City infrastructure layout and land use

Years 0 100 200 300

Replacement rate of different elements of the built environment

believed to cause numerous health problems, including suppression of the immune system.

The designer can circumvent some of these problems. For example, lead-free water-based latex paints might be employed, and wool/nylon blend, formaldehyde-free carpets can be used over 100 percent-natural jutepad with homosote underlayment. The furniture used in our building needs to be tested by an independent laboratory for the presence of formaldehyde.

The designer should ensure that the selection of materials and systems also avoids the use of ozone-destroying chlorofluorocarbons (CFCs) and hydrochlorofluorocarbons (HCFCs). About 50 percent of all CFCs used are in buildings. The main measures to phase out CFC and HCFC uses are:

● designing buildings to avoid employing air conditioning that use these;

● avoiding specifying insulation materials that use these in their productions;

● avoiding halon-based fire-control agents;

● designing buildings to maximise natural light and ventilation;

● upgrading existing CFC air-conditioning systems with alternate systems, and avoiding the repeated use of CFC-related materials including HCFCs.

Another hazardous material is asbestos. The designer must ensure that the building is constructed without the use of asbestos or asbestos-containing material. Existing skyscrapers that are being adapted must have an asbestos operations and management plan. VOC content should be limited in adhesives, architectural sealants (e.g., used as fillers), paints and coatings.

● The Availability of the Material in the Biosphere
The general principles of resource management can provide a
means of resource conservation, as can more conventional
approaches. The natural resources that occur in the biosphere can
be classified under a variety of rubrics. For example, resources can
be grouped according to their sources of origin: forest products;
non-metallic minerals and those products made from them; products of a single metal; and miscellaneous products and compound
products. It must be understood that each building element found
in the built environment possesses its own history of energy and
materials consumption. Every part of a building or structure can
produce pollution and ecosystem degradation. Producing a particular building element and making it available for use in a building
project necessarily entails environmental consequences.

'Natural resources' are considered to include energy sources as
well as materials extracted from the earth and exploited by
humans. These are further categorised according to their availability and potential for regeneration (Skinner, 1969; Flawn, 1970; Common, 1973). The distinction between renewable and non-renewable
resources is the 'appreciation of their relative importance in relation to the variables which constitutes the environment'. Some hold
that this distinction is basic to the ecological approach to conserv-
ing resources (Costin, 1959). The external dependencies of the built
environment, and the resources on which it depends, are categorised as follows:

● Inexhaustible Resources
Examples of these sorts of resources include air, water and solar
energy. Although the total available quantity of each of these
resources is virtually unlimited, the form in which each occurs is
subject to change. Such variation has particular consequences for
the ability of these resources to sustain life. Any permanent degradation of their quality (for instance, as a result of pollution) therefore raises concern.

● Replaceable and Maintainable Resources
Examples of these resources include water, as well as flora and
fauna. Simply put, replaceable and maintainable resources are
those resources the production of which is primarily a function of
the environment. In environmental conditions such resources will
be produced indefinitely. However, any impairment of the environment will have an adverse effect on the production of these sorts of
resources. The long-term viability of this category of resources
depends upon many factors, but especially on the deliberate and

inadvertent interference of human beings in the production of these resources.

● Irreplaceable Resources
Examples of such resources are minerals, soil, fossil fuels, land and the landscape itself in its original state. These resources are known to be irreplaceable and their availability is related to the rate and type of exploitation by humans. Non-renewable energy resources are essentially the results of solar energy in the past; therefore, they exist in a finite amount. Present consumption of these resources by humans is at a rate that makes natural regeneration rates negligible. This situation raises an acute multi-generational allocation problem: the more of these non-renewable resources we use now, the less will exist in the future.

This categorisation of resources clearly indicates which of them demand our conservation efforts. The categories are flexible as new substitutes are found for traditional resources or as extraction and recovery methods evolve (O'Riordan, 1971). Both developments – resource replacement and discovery of extraction techniques – have ramifications for supplies. Because building materials made from naturally occurring resources are used worldwide, the use of such materials, as well as of energy resources, in the building will have ramifications for the future availability of resources.

Common mineral and energy resources that are irreplaceable can be classified further, based on use (Skinner, 1969):

● Non-Renewable (Metallic Mineral) Resources
These include abundant metals such as iron, aluminium, chromium, manganese, titanium and magnesium. Scarce metals are also among these non-renewable resources. They include copper, lead, zinc, tin, tungsten, gold, silver, platinum, uranium, mercury and molybdenum.

● Non-Renewable (Nonmetallic Mineral) Resources
These also include minerals used in chemical fertilizers and for special uses: sodium, chloride, phosphates, nitrates, sulphur. Some of these are mainly building materials: cement, sand, gravel, gypsum, asbestos and so on. Water is also a non-renewable resource and exists in lakes, rivers and groundwater.

● Limited and Non-Renewable (Energy) Resources
These resources include fossil fuels, e.g., coal, oil, natural gas, and oil shale, as well as resource material suitable for nuclear fission or fusion.

● Continuous-flow (Energy) Resources. These include primary and secondary forms. Examples of direct use include: the flow of precipitated water, tidal effects of water, geothermal phenomena, wind power and climate energy. Examples of indirect use (for example, through combustion) include: photosynthesised energy (e.g., wood) and waste products used as fuel.

These categories can be useful to the designer since they remind him or her of which are the renewable and which the non-renewable resources. The status of a resource should be taken into consideration in design and in the management of inputs and outputs. The earth contains finite quantities of the non-renewable materials and energy resources. This ecological truth cannot be disputed. However, what does spark disagreement between resource conservationists are the quantitative questions: in what quantities do resources exist, where are they located and how much of them can be extracted over what period of time? (e.g., McHale, 1970; Kahn, 1978). The earth and the biosphere are closed materials systems, and consequently the present pace of continuous and accelerating human consumption cannot be sustained indefinitely. Available mineral deposits, including fuels and metals, were formed over the course of geological time, but they are being consumed far more rapidly than they could possibly be regenerated. As the increasing demand for minerals and fuel sources leads to the use of lower-grade resources, the problem of resource depletion is temporarily resolved, but this resolution only comes at the expense of increasing pollution and ecosystem destruction.

Ecological design has to consider the human and technological parameters of non-renewable resources rather than those determined by geology. Some resources, such as precious metals, are depleted less for their use value or demand and more because of their economic value. The high market value of such substances provides a motivation in itself for their extraction – though also for their reuse.

Energy and Non-Renewable Fossil Fuel Resources

The above discussion has demonstrated that the ecologically sensitive designer must acknowledge that a consequence of the use of any element in a built system is its history of direct and indirect interactions with ecosystems. The designer must recognise that each choice of building material has an impact on the earth's resources. We can analyse this impact by using the interaction framework. Every activity that is part of producing an element for incorporation into a built system will, first of all, exert spatial pressure in its immediate surroundings and will have an impact on local ecosystems. Every building project also affects other interventions in the built environment. A building also relies on consumed energy and materials for its operation. Moreover, any part of the built environment produces outputs and waste energy which may be expelled into the ecosystem and contaminate the environment. These comprise the sum of the ecological consequences of the exchanges of energy and materials between the natural and built environments (i.e. L21).

In order to assess the extent of ecological impact, one must consider a number of related factors. First, the designer must take into account the entirety of processes and activities required to make each material or energy resource available for use in a designed system (as discussed in chapter 4). Second, the displacement of natural areas by these activities and processes and their impact on natural systems in their immediate contexts are also considerations (see chapter 5). Third, the total amount of energy and materials required by these processes and activities, and the respective impacts of such energy and materials on the ecosystems, must be analysed (dealt with in chapter 4 as well as in the present chapter). Fourth, the total output of energy and materials from each of these processes and activities, and their environmental costs to ecosystems, has to be considered (see chapter 4).

These varieties of environmental consequences are, of course, interrelated as demonstrated in the partitioned matrix in chapter 3. Their net impact constitutes the total ecological cost of using given energy and material resources to construct a building or group of buildings and related structures. In a design approach that takes seriously the environmental ramifications of building, a complete analysis of the ecological consequences of the built environment's components will undoubtedly become a time-consuming and complex task. It will entail an analysis of the ecological impact of using each material and energy resource employed in the built structure. This analysis becomes even more burdensome when it is remembered that each resource has to be viewed from its place of origin in the environment to its ultimate re-absorption (following demolition of the building or complex) into the biosphere. However, indicators can be used to facilitate the analysis.

Of critical concern is the amount of energy used in built systems. Current accepted patterns of exploitation of material resources depend largely on non-renewable fossil fuels as energy sources. These non-renewable resources are constantly becoming

more scarce; in many cases, they can be recovered only by additional expenditures of energy, which in turn must be procured somewhere in the environment and thus cause additional environmental impact.

At present rates of use of
non-renewable energy resources, within the next
50 years the world will likely run out of
non-renewable fossil fuels

(unless new sources of these fuels are discovered). It is then imperative to use renewable sources as much as possible (e.g. solar energy,
wind energy, geothermal, etcetera). Wind energy is proving to be a
viable option where local climatic conditions permit (i.e. average
wind speeds of about 6 metres per second or more). Currently wind
generators have efficiencies of 40–50 percent, with an upper theoretical limit of efficiency at 59.3 percent (O'Connor, 1996, in
Edwards, 1996). Other alternatives include biomass (in Zeiher, 1996),
photovoltaics and other sources, all of which deserve further development.

In general, major existing building developments depend upon
fossil fuels and have even been designed to be extravagant in
their use of these non-renewable resources. This situation has
resulted from the fact that, until recently, the built environment
has seldom been considered in its totality from the perspective of
energy consumption. Considerable amounts of energy are used to
produce a building and to maintain it thereafter. Energy is consumed to operate a building, to move people and goods to and
from it, and to demolish it when it is no longer needed. Buildings
also exert a major impact on the flow of energy in their local area.
Moreover, even the process of coping with the environmental
problems brought about by buildings requires expenditures of
energy (for instance, recycling solid wastes and reducing air pollution are not 'neutral' activities, but require energy). Buildings and
other associated structures consume about 30 to 40 percent of the
energy required by the manmade environment in a given year,
and transportation takes up another 25 percent (Bender, 1973).
Increasing the supply of energy will be environmentally counter-
productive, in view of the fact that greater quantities will result in
increased use and in the concentration of materials – thus, only
encouraging resource consumption. Such an increase would only

exacerbate current energy problems by facilitating existing patterns.

The availability of non-renewable energy resources also has an impact on the availability of other material resources, and ultimately on the ability of the built environment to function. By using fossil fuels, we increase the earth's environmental difficulties. Clearly, as designers we must work to reduce overall energy consumption, which will lead to positive environmental consequences.

In order to assess the value of a designed system or the cost of a service, we add up the quantities of non-renewable energy resources that were required to produce the system or service (e.g., Berdurski, 1973; Hannon, 1973; Howard, 1997, et al.). The net result of this accounting is the embodied energy value of the material. Certain products consume more energy per unit than others. If we consider the fact that any resource use has an unavoidable impact on the environment, and that most energy resources used in the existing manmade environment have been produced from non-renewable sources, then the ecological/environmental impact of the components of the built system can be accounted for during the design process by using a form of 'energy equivalents' (e.g., Makhijani and Lichtenberg, 1972; Hirst, 1973; Chapman, 1973, 1974; Howard, 1997, et al.).

The energy cost of a designed system must be considered in the largest sense, that is over the course of its entire life cycle. The energy that goes into the operation of a built system during its life is great (around 65 percent) in comparison with the energy used in its initial construction. More dramatic is the finding that for every kilowatt hour of energy consumed in the construction of a commercial skyscraper (including energy used in the manufacture of materials, their transportation and so on), an equal amount of energy will be used to operate the building each year over the course of its useful life (GSA, 1974).

Within the field of ecology, energetics is the study of the ecosystem with respect to energy exchanges and metabolism (or efficiencies). Energetics converts the biomass of an ecosystem into energy and units. Energy passes through a built system, and this energy can be quantified using indices, which allow us to see the built environment in terms of the energy cost of producing building materials (Beckman and Weidt, 1973; Chapman, 1973; Yeang, 1974a, Howard, 1997, et al.). The energy cost of a product provides an indication of its impact on the environment through the use of energy resources, and also lets us compare its cost to that of other products. This method can be used to assess the efficiency of current technology by considering together all the energy used in a product's creation, operation in the built environment and later disposal or recovery. Hence the total energy costs of different construction processes and materials can be compared (e.g., Brown and Stetlon, 1974) and their relative degrees of dependence on the earth's resources can be assessed.

There are a number of factors to take into account when selecting materials. If we are approaching design with concern for environmental consequences, we must thoroughly analyse and quantify the energy and material resource requirements of a building over the course of its useful life, and we must inventory the ways in which the materials and resources impair the ecosystem (that is, we must determine their 'embodied ecological impact') (Patterson, 1990; Roaf and Hancock, 1992). Designers must also go beyond functionalism in selecting materials and energy forms for use in their projects. One design criterion should be the environmental impact of a particular energy source or material over the entire life of the building. When thinking about materials, it will be important for designers to consider whether a material is a renewable or a non-renewable resource. This assessment is actually complex, because each process in the built environment requires resources and exerts a particular impact on the external environment.

The purposes of incorporating these energy and materials considerations into design are several. First, the designer works to reduce the depletion of the fund of resources in the biosphere. Second, he or she attempts to minimise, through good design, the outputs expelled by the built environment into the ecosystem. Third, the designer works to reduce the built environment's use of resources from the natural world.

By analysing the exchanges between the built environment and its ecosystems, the designer amasses information, which is useful in considering aspects of a proposed building project. Such information supports a comparison of the total consequences of alternative design schemes with respect to the use of materials and energy. This comparison will in turn inform the choice of sub-systems in the chosen design. The designer can also configure the project in such a way as to conserve non-renewable energy and materials. He or she can determine which aspects of the built environment require excessive quantities of resources and then target those profligate energy consumers that can be modified to achieve a more reasonable level of energy consumption. The objective of minimising environmental impact can be met by planning for efficient energy use and minimum resource consumption. Reducing energy use will in turn lead to a reduction of ecosystem effects – always keeping in mind that the building's impact profile takes in the whole of its useful life. Thus, designers and their clients can plan for resources to be recycled or recovered rather than simply wasted at the end of the production and operational phases of the designed system. Such strategies as lowering supply and flow rates, increasing efficiency and performance of systems and the designing new systems all contribute to resource conservation.

It must be observed, however, that a decrease in the use of high ecological-impact materials and forms of energy as a result of the efficient use of alternative materials would only reduce to some extent the overall impact on the biosphere. The best long-term solution is to reduce overall demand through the modification of human patterns of needs (i.e., reduction in standard of living) and through the general practice of conservation in the use of material and energy resources (see chapter 4). As has been mentioned, this optimal strategy also has larger social, economic and political ramifications that lie outside the scope of this work.

If we consider where in the entire life cycle of a building to intervene on behalf of the environment, it is clear that the design stage is the crucial point. It is at that moment that many decisions are made that have a profound impact on the amount of energy and materials used over the course of the building's whole life. The original designer is the individual who initially specifies the types and qualities of the materials that will go into the building, albeit with the input of other participants in the process. The processing or fabrication methods that the designer chooses or advocates for the manufacture and recovery of a product will have enormous consequences for the amount of waste that will be eventually generated in the building process and beyond. Hence, the decisions that we make early in the design process are of great importance. For example,

if a designer makes a poor choice of a certain material, its failure will require replacement. In this scenario, the amount of material used is doubled, as is the quantity of resources consumed. Even more dire environmental consequences result in the case of a failed product that is an inseparable part of a larger assembly which is in turn rendered useless and thus expelled into the environment as waste (Ballard, 1974).

In the area of materials processing, it is important to select fabrication methods that do not require excessive machinery or result in excessive waste (or both). Such processes will increase the total amount of materials and energy resources consumed per unit of building product. The use of materials, forms of construction, production technology, transport, assembly and dismantling of building components must all be planned in relation to energy content and ecological impact and the life cycle of materials.

Using an environmentally sensitive approach, the designer thinks in terms of efficiency and reduces the total amount of resources required. The consumption of energy and materials thus decreased, the overall flow of energy and materials through the built environment lessens. The designer helps ensure that the processes built into the design of the building require the smallest quantity of materials and energy, choosing among the alternatives capable of performing a given service. In meeting this objective, the designer has to do more with less (Fuller, 1963). Beyond working towards a rational use of energy and materials in the initial construction of a part of the built environment, based on resource conservation principles, the designer also takes into consideration the projected reuse and recycling of the selected elements.

Designing the Physical Life of the Building's Components

In ecological design, we must guarantee – as much as possible – the reuse or recovery of materials. In the selection of materials in the green skyscraper or similarly large buildings, we must also consider the building and its components as having overall physical life spans of perhaps 50 to 80 years, with subsystems (for example the cladding) having lesser life spans (about 5–10 years), along with M&E equipment (e.g., 10 to 15 years). Different materials, components and equipment have their own life spans, after which replacement and recovery need to be implemented (see Fig. 44).

The building's physical life is also different from its economic life. The economic life of a commercial building is considered to be that period during which it produces a financial return that justifies the investment made. In the case of an owner-occupied building, as opposed to a speculative project, its commercial life is that period during which it is used directly (Weimer and Hoyt, 1966). The economic life of a building is thus distinct from its physical life, which is potentially much longer. There is, nonetheless, a correlation between the physical life and the period in which the owner of a building receives an adequate return on investment.

Fig. 44   Production phase
in the life cycle of
a single building element
(a metal) (Source: Yeang,
1995)

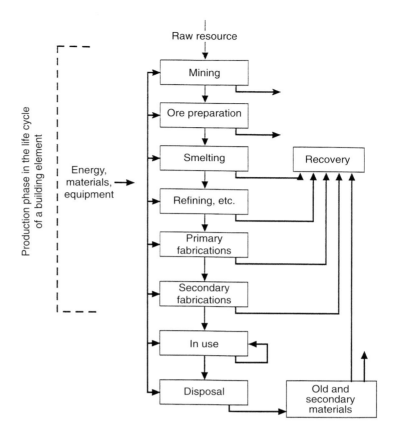

The ecological physical life spans of components of a built system are potentially very long. The design professions today, however, are not constituted in such a way as to emphasise considerations of the physical lives of building components. Rather, the
economic lives of buildings and their parts are the primary considerations in most projects. However, when we adopt the ecological
approach, we necessarily concern ourselves primarily with the
ecological lives of our built works.

Under the existing system, which emphasises economic considerations, real estate financing, as well as the design process, is
based on the expectation that commercial buildings will last
around 30 years. This is the extent of the project's economic life,
after which it is deemed worthless (Crosby, 1973, p.9). Obviously,
most standing commercial buildings, by virtue of their construction
out of durable materials, are capable of outlasting the relatively
brief economic life imagined during the investment process. The
'throwaway' culture of much of contemporary building and design
practice results in buildings that are uneconomical and unecologi-

cal white elephants. Beyond the 30-year mark, these buildings are rendered nearly obsolete: because their long physical lives were not taken into consideration during the design process, they have become difficult to reuse or renew.

There are a number of factors that may contribute to the obsolescence of an entire development project, building or building component. Here we can speak of varieties of obsolescence related to different causes, some internal and others external to the structure's design. Factors that contribute to this process of loss of value include location obsolescence, which is to say that the building's original function may no longer be appropriate in its immediate vicinity, or the function may no longer be required as a consequence of social and economic forces (e.g., the rendering obsolete of the government buildings in Bonn after the removal of the German capital to Berlin). Developing technologies also pose challenges to existing buildings, since many may not be capable of adapting to such changes. The forces of nature themselves may also make a built system, or part of it, obsolete, since they cause wear in structures that may push them below established standards of comfort and safety. Finally, changing statutory regulations, such as building codes, may lead to the obsolescence of buildings.

Very often, a value judgement needs to be made about the anticipated life cycle of the building. For instance, aluminium has a higher embodied energy than steel; however, at the end of its useful life in a building, it requires considerably less energy to recycle than does steel. Making aluminium from recycled aluminium uses 90 percent less energy than making it from scratch, and cuts related air pollution by 95 percent. Similarly, using recycled glass for glass manufacture saves up to 32 percent of total energy required, reduces air pollution by 20 percent and water pollution by 50 percent. I have gone into the outputs of the built environment and their possible routes in greater detail elsewhere (Yeang, 1995). A diagram of the materials flow from 'cradle-to-grave' will be discussed later in the chapter.

We should note that the largest component in the skyscraper and other similarly large intensive building types is its structure. It follows that the most fundamental step will be the selection of a structural system that supports the floors and fabric of the building. Selection will be influenced by factors such as recyclability of material; ease of access to the site for delivery of building components such as structural steel, precast components or ready-mixed concrete; the disruption construction procedures may have on surrounding urban activity; noise restrictions imposed by authorities; time constraints; and the final cost.

Life Cycle Considerations

Like a living organism, the operation of a present-day skyscraper and other similarly large building types requires a constant feed of materials and energy resources and, at the same time, the process generates continual outputs. The process of extracting the necessary resources from the environment, processing them, and delivering them to the building where they are consumed exerts a significant impact on ecosystems and depletes the earth's limited fund of non-renewable energy and materials. The designer of a skyscraper and other such intensive buildings must consider the anticipated resource needs of the building early on. This is particularly the case when we consider that each material and energy resource poured into the built environment necessarily brings about spatial alterations to the ecosystems as well as a depletion of the total volume of the resource that exists in the environment.

Analogising the building with the ecosystem, we can categorise the components of a typical skyscraper and other intensive buildings as biotic and abiotic elements of the earth, which have been processed or assembled piece-by-piece into the built environment. By its process of ecological bookkeeping, the green design of the built environment becomes a kind of resource management. In this process of 'management by design', each part of a building is accounted for and projected to the moment of its eventual reuse or recycling. Instead of being regarded, as it is traditionally, as a fixed and immutable quantity, the built system (whether a skyscraper or other building type) might be better regarded as a mass of materials contained temporarily in the form 'building' for a brief period of use, after which the components return to the continuous flow and exchange of energy and materials within the biosphere. This 'ashes to ashes' way of considering the built environment is analogous to the method used by ecologists to view the exchanges of energy and materials within an ecosystem (e.g., Odum, E. P., 1963).

From the perspective of materials and energy management, a skyscraper, or any building for that matter, represents just one moment in the lives of its component parts. The skyscraper and other intensive building types is a built form in which people have assembled a quantity of energy and matter, employing certain already established patterns of assembly, and projecting a particular use pattern. This assemblage is then imposed on the ecosystem of the project site. When it eventually becomes impossible, uneconomical or unfeasible to continue to use the building (due to the various forces of obsolescence outlined above), the component materials may well be removed and then either reused elsewhere in the built environment, recycled or simply disposed of (for example as landfill). If the materials are biodegradable, they can be absorbed by the natural environment within these limitations, and if not, they nevertheless remain present as pollution.

Taking the broadest possible view, we can conclude that designed systems and their component parts constitute just one short moment in the far longer cycles of the biosphere's ongoing patterns of energy and materials exchange. If the designer sees the built environment thus, he or she is led to view it synoptically as part of the greater biosphere on which it depends for its existence. The designer must therefore account for the quantities of materials and energy that the designed system uses and expels as a part of the larger manmade environment. The designer must also account for the impact on the environment that will result from this resource use. For example, global building construction generates about 40 million tons of carbon dioxide and lesser quantities of acid gas and nitrous oxide per annum. Taking these outputs into consideration in the design process would be a step toward understanding the interdependence of the built and natural environments, which is at the heart of environmentally conscious design.

In assessing ecosystem impairment, we will want to attend particularly to the transfer points at which it ordinarily occurs. At these sites, inefficient technologies and bad designs can cause gross impairment of the ecosystem. One such transfer point would be the place where raw materials are extracted from the earth (for example strip mining). Transportation, construction and recovery processes can degrade the environment. The end point or 'sink' of the material and waste energy is also an obvious transfer point where damage occurs. The extent of such damage will depend on the sort of energy or material that has been used. For instance, in comparison to iron, aluminium requires more energy for its production and creates more environmental harm.

The designer's concern is with the extent and quality of human use of the ecosystem and the earth's resources in the system he or she is planning. But the designer must also consider the ways in which the elements of the building are extracted from the earth and warehoused before their assembly into the built system. Thus, green design imposes on the architect a larger horizon than traditional design, in which materials are treated as if they were ecologically neutral and limitless in abundance. The lives of the materials in the built structure will be of concern to the designer, as will their ultimate disassembly, followed by what will hopefully be a low-impact or even beneficial reintroduction into the biosphere (see Steele, 1997). The designer should therefore design with end use in mind. It has been estimated that there is potential for 75 percent of all construction waste to be reused or recycled. Currently in Europe, only 5 percent of building materials are recycled.

At many moments in the life of a designed system, a given quantity of energy or a particular material has the possibility of interacting with the ecosystem. Mining and extracting minerals from the earth causes widespread habitat destruction. Such processes devour vast quantities of energy and other material resources, while equivalent amounts of waste are generated, including greenhouse gases, hydrocarbons, hazardous chemicals and solid wastes. These waste products are also generated in processing the materials and transforming them into products that can be incorporated into a building. The inputs of materials to our skyscraper and other intensive building types occur at all stages of production, construction, operation and recovery.

The transfer points that are so crucial to understanding environmental impact are part of the long route taken by each material resource, from its environmental position prior to use to its final resting place. In principle, the designer is obligated to consider this entire trajectory.

The environmental consequence of using a given resource goes back to the process by which it was made available for use; the early stage of the life of a material is not usually considered by the designer who is trained to think only about the assembly of components into a structure located at a particular project site. What designer, for example, is encouraged to think about the fact that mining and other extraction techniques associated with the metals used in architectural projects can lead to the destruction of natural habitats? Yet designers should know that when rocks containing minerals are removed from a mine, the waste rocks are customarily left somewhere nearby in the form of tailings or open dumps. Rising demand draws down stocks of the mineral and pushes production toward lower-grade ores; the land area on which the extraction process has an impact increases, while the quantity of waste rock rises. Not only are large amounts of energy and materials used to mine, transport and process materials for building, but an equally large quantity of waste materials and energy are generated in the process. For example, the slag, or solid waste products, as well as waste gases, derived from the refining process have to be ejected into the environment. And when the mineral resources within the mine have been exhausted, we are left with a degraded landscape that is full of holes, mountains of rock and scattered buildings and machinery now made obsolete (LaPorte et al., 1972). The open gouges in the earth's surface may make toxic or harmful substances available to groundwater, producing dangerous runoff for decades after the mine has been abandoned. The cycle of waste, damage and energyconsuming processes continues to expand almost limitlessly. To reclaim a seriously damaged landscape requires yet more energy and material resources. Beyond the mine site, the transportation of its products, their use in the built environment and their eventual disposal eat up even more energy and materials and produce yet further waste products.

Figure 45 shows in schematic form the production phase, which is part of the longer life cycle of a building component. In each of the compartments, the element interacts in some way with the ecology, and each of the processes in a compartment exerts some degree of impact on the ecosystem. Furthermore, every element requires energy and materials in order to function. Let us not forget that no process takes place in a vacuum: each requires plants and machinery, building enclosures and a supply of energy and materials to operate. For instance, the complex set of inputs to enable one process (e.g., mining) to take place is shown in figure 44.

The character of the interaction between the built environment and the natural world varies with the stages in the life cycle of a building. In the production phase, raw materials and energy are

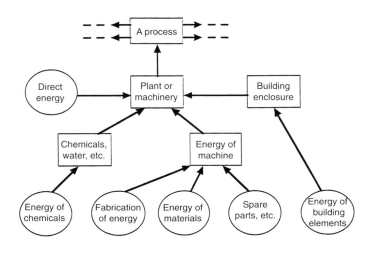

Fig. 45   The primary inputs to a single process in the built environment (Source: Yeang)

Fig. 46   Hi-Rise Recycling
Systems, Miami, Florida
(Source: Hi-Rise Recycling
Systems, Miami, Florida)

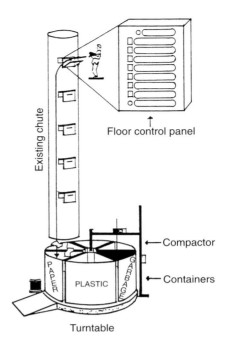

Recycling systems can be retrofitted or specified in buildings to make recycling
easy for building occupants. One simply pushes the button specifying the material;
a green light and buzzer indicate that the correct bin is under the chute, and the
door unlocks so the trash can be thrown away. The recyclable material travels
down the chute and lands in the appropriate bin below. The bins are placed on a
turntable that shifts in order to accommodate several types of materials.

extracted from the environment, prepared for use and distributed
to building sites. Next comes the construction phase, when spatial
interaction with the environment on the project site begins. At this
point, building elements are fabricated and assembled. The various
processes entailed in construction will all consume materials and
energy. Once construction of the new component of the built environment is finished, the operational or consumption phase begins.
The processes witnessed at this phase include the functioning of
the building or complex, which will inevitably require maintenance
and modification in response to the needs of users. The later recovery phase constitutes the coming full circle of the life cycles of
materials and energy resources. This phase can include one or more
of a number of alternative ways of treating the built work. If the
existing building can be effectively retrofitted, it can be renewed or
reused. In cases where a building is irredeemably obsolete, it may
be removed or demolished. Some of its component parts may be

salvaged. At this point the site itself may also become the focus of rehabilitation efforts, for instance by replanting native plants and reintroducing indigenous wildlife.

Thinking of the built environment in terms of these cyclical phases, we can identify the ecological interdependencies between the components of the built environment as well as assess the environmental impact of each. This is the only legitimate way of determining actual environmental cost, and the most credible means by which the claims of entire buildings or individual products to 'green' status can be evaluated. Life cycle inventory (LCI) provides a quantitative assessment of environmental inputs and outputs associated with a product (Steele, op. cit.). This model illustrates possible life cycles for energy and materials. At each phase in the cycles, we observe materials and energy being taken in, and other materials and forms of energy being released as outputs. It is important to see that these processes unfold on a particular location, which may sustain a negative environmental impact as a consequence of such activity. We can test our findings with the interactions framework (see chapter 3).

By looking at the built environment from this perspective, we are led to consider energy and materials use well beyond the consumption phase, which is our usual focus as designers. It is easy to see the environmental consequences of building, but the processes by which the individual components were produced are often overlooked. Each component in a building project represents a certain amount of energy and material consumed, as well as an amount of pollutants emitted and a portion of ecosystems degraded. If the ecological designer truly wants to be comprehensive in his or her assessment of environmental impact, he or she must consider every element in the building from its source in raw materials and make an accounting of all the material and energy used, as well as the volume of waste produced. The designer must also evaluate the total impact on the natural environment made by each element in the building throughout its entire life cycle. Taken individually, the stages in the life cycle may only represent minor environmental harm. But when the designer looks at them collectively, inappropriate levels of environmental impact may emerge, such as excessive waste of resources and degradation of the environment out of all proportion to the scale of the project. As the architectural profession and related disciplines are presently constituted, and as the development process is structured, the life cycle of a building project is only viewed from an economic perspective. Decisions about design and building that make good economic sense often do not make good environmental sense.

Although I am arguing for ecological bookkeeping – indeed, a balanced environmental budget – to occupy an important place in the design and development processes, one has to be realistic. Designers should not attempt to create a closed system in which all materials and energy are entirely internally recirculated and reused. Not only would such a state of equilibrium be impossible to achieve, it might not even be desirable for all cases.

Linear and Cyclical Patterns of Use of Materials

There are two essential patterns of resource use: the linear and the cyclical. The linear pattern is favored in the industrialised world, where non-renewable resources flow once through the built environment where they are used then thrown away. In effect, this is a transformative process by which resources are converted into wastes. In between these two end points in the process there is a relatively brief period in which the resources are actually put to use by human beings (Davoll, 1972, p. 335). The end result of this linear approach is a waste problem, centered on the issues of disposal and environmental contamination. Only conservation can stem the waste of increasingly scarce non-renewable resources by this all too common pattern of usage.

One way that conservation could extend the life of a processed material would be through systematic recovery throughout the built environment. For example, the building industry now wastes metallic minerals when parts of structures are used only once and then discarded. If we think of those metal parts in the broad terms outlined above, we see that they represent far more than the material in the form of a pipe, column or other fitting. The metals from which these pieces of the building are made had to be located in the earth, extracted, transformed into building materials and transported to the site, all at the cost of energy use and environmental degradation. When we discard an element of a building, not only do we essentially waste all of the energy that went into its making, as well as rendering the environmental impairment its manufacture brought about entirely useless; in addition, we also create the need for a new product. Thus new demand is fed into the linear conveyor-belt system of resource exploitation, for the production of a new element (similar to the one that was thrown away) will entail the use of more resources and require additional energy expenditures. By discarding elements we also create disposal problems. This is not to say, however, that by just extending the life of a built system we will solve disposal problems; instead, we may only delay them. Nor will the substitution of more durable products necessarily solve the problem, as better quality elements may cause even greater environmental impact than poor quality materials.

Analogies with the natural cycles of growth and decay as temporally opposite points may be useful in conceiving of the cycling of materials through the built environment. For example, we could think of cities as occupying various points along this continuum. Thus we could understand urban renewal projects as attempts to rectify the decay of existing building stock in older cities. One consequence of decay is the disposal of elements of the built environment, which are often dumped without first being separated. Nonetheless, a certain amount of salvaging of building materials is already being undertaken. Certain parts of buildings are recovered as scrap; for example, bricks and rubble are sometimes reused as landfill. It has been estimated that perhaps as much as a quarter of the volume of material resources consumed by building is recycled in the built environment (Klaff, 1973). By thinking of the inevitable final disposition of the materials consumed by a building project, the designer can remedy the prevailing emphasis on the first costs of buildings. Alternatives for reusing elements of buildings, which must necessarily someday be rendered obsolete and demolished, can be figured into a broad assessment of energy and materials use over the life cycle of a designed system. Of course, a conceptual shift is required for designers to think that their buildings are not permanent structures. It is perhaps useful to view the cycle as a series of transformations (Ashby, 1956). If, according to this model, a sub-system has four obvious stages, a, b, c, and d, and the transformation goes a-b-c-d-a-b and so forth, then it can be said that these transformations are cyclical (Schudltz, 1969). The materials in the built environment could be similarly transformed.

An example of a rating system used in the USA for green buildings suggests ratings for buildings in which 20 percent of the total building materials by cost are from manufacturers located within a 300-mile radius of the building site, for buildings that use salvage or refurbished materials for 5 percent of total materials by costs, and for buildings that use a minimum of 50 percent of the materials as measured by cost (exclusive of cost for mechanical, electrical and plumbing systems) that contain at least 20 percent post-consumer recycled content, for developing a construction and waste management plan that uses licensed haulers of recyclables and that documents costs for recycling and frequency of pick-ups (e.g. of cardboards, metals, concrete, brick, asphalt, land-clearing debris and beverage containers).

If the recovery of all manmade products were built into the design process by insisting upon kinds of reuse, regeneration and recycling that require minimal amounts of energy, the overall reliance of the built environment on non-renewable resources could be reduced. The need for non-renewable resources would also decrease as a consequence. Given the fact that the quantities of fossil fuels and other materials in the earth are limited, our ability to recover materials from the obsolete portions of the built environment must increase. We must also minimise the degree to which we waste resources and we must limit our use of resources that cannot be recovered after use. Some quantities of what we consider to be 'pollution' are in fact accumulations of resources in inappropriate locations, or derelict resources occupying non-functional spaces which could be rehabilitated and put back into the manmade system. We would not think of such materials as pollution had they been recovered and put to new uses.

There are a number of ecological benefits that would arise from a cyclical pattern of use of materials in the green skyscraper.

We could reduce the amount of pollution generated by the building, conserve the amount of resources consumed, decrease the quantity of potentially harmful materials to be disposed of and contribute to an overall reduction in the use of energy and material resources by the built environment. Ideally, in a cyclical pattern of resource use, a particular material or energy source would be extracted from the environment, used in the building, then transformed in an energy-efficient manner into a new resource. The process would then become a closed loop. A process that had formerly culminated in the production of a discarded material (waste output) would now end in the production of another resource (productive input). Our ultimate objective, therefore, is to make more cyclical a process that already possesses that characteristic to some degree (Odum, E.P., 1971).

A familiar cautionary note should, however, be sounded again: it will likely never be possible to achieve a hermetic cyclical pattern of resource use in the built environment. The recovery cycle is one stage where there will probably always be a quantity of resources lost. For instance, in the process of recycling scrap steel, approximately 10 percent of the material is lost. Some materials cannot be put back into the system after initial use at all: inherent in the use of products like paints, solvents and cleaners is their dispersal in the environment. Other factors also contribute to the loss of material in the system, including friction and oxidation. Furthermore, thermal emission would represent a loss of energy even in an ideal system in which every element in the built environment was reused or recycled.

In order to minimise the environmental costs of reuse and re-cycling, this phase of the cycle of resource use must be addressed at the design stage. It must be understood that incorporating resource recovery into the built environment from the outset will necessarily lead to additional resource use and environmental costs. The point is to ensure that the recovery phase exerts the smallest possible impact on the ecosystems and upon the use of resources.

There is an argument that design should use materials close to the building site so as to reduce the cost of energy used in transporting the materials. The issue is more complex than simply saving on first costs of energy attributed to a project. Some materials (e.g. steel), although incurring high transportation costs to some locations, can be reused or recycled many times after first use. There is an argument that, because of the abundance of timber, that timber should be the predominant material for construction. But typically hardwood demand is met by forestry practices that are not sustainable. A few companies claim to replant, but the vast majority of felled timber results in the loss of irreplaceable rain forests. Many species have become extinct as a result, and many more are now endangered. The simple design decision is not to specify any tropical woods.

Subscribing to the cyclic pattern of use of materials means that the reuse or recycling potential of a material far exceeds the benefit of selection based on proximity of materials or lower embodied energy.

Designing for Recovery

Embodied in the building are the results of every decision made by its designer regarding projected uses of materials and energy. Ideally, as we have argued above, the designer should work to achieve a cyclical pattern of resource use in the portion of the built environment for which he or she is responsible. The possibility of achieving a cyclical pattern of resource use in a building or larger development project is dependent on a number of factors. For example, the costs in terms of energy and materials of establishing a cyclical use pattern have to be taken into consideration. In current development practice, the economics and efficiency of constructing buildings and of later maintaining them is an important consideration. However, the amount of energy required to dismantle the building is rarely, if ever, considered. In the minds of architects and developers, buildings are built to be permanent despite the fact that the economic lives of their components is becoming increasingly short (McHale, 1967, p. 123).

In the process of green design, we must consider the amounts of resources that will be used to dismantle the building and its components, and we should consider as well the amount of pollution and waste that will be generated in the process. As we have already observed, recycling building materials requires further energy expenditures. Thus, our choice of materials must be informed by a consideration of the relative environmental costs of recovering and reusing them in some form. When we consider the environmental costs of reuse, we must look not only at resource use but also at the impact that the process will exert on the ecosystem. Dismantling the portion of the built environment and recovering some or all of its components will necessarily have ecological consequences. For example, demolition may entail the use of heavy equipment that cannot be transported to a building site, set up and operated without creating some kind of impact on the immediate surroundings.

It may even become necessary in some cases to erect processing facilities to process the demolished materials on site, which will also create a stress on the local environment. Moreover, the recovery process may lead to emissions of various materials, possibly in the form of pollution. The designer must weigh his or her objective of cyclical resource use, as well as his or her choice of recovery process, against the qualities and quantities of emissions likely to be generated.

The availability of the raw materials will also inform our choices. The economic viability and ecological necessity of resource recovery is also a function of how rare or abundant a particular material is in the environment, and how easily it can be extracted for human use. Gold, for example, is regularly recovered because of its high economic value. Aluminium, on the other hand, is recovered not because of its rarity – it is in fact abundant – but because producing it from raw resources is costly in terms of energy while reprocessing it is relatively less expensive.

The possibility for recovery after initial use will be influenced by the form of construction that has been employed in the obsolete building or buildings. For instance, it is relatively easy and cost effective to collect steel and other metals from a building and melt down the scrap. Reinforced concrete, by way of contrast, is a building material produced by a chemical reaction and it cannot be separated into its component elements of sand, cement and steel. Its reuse is limited to employing material in its downgraded form as landfill or hard-core and in certain instances as re-used concrete aggregate. Clearly, building systems produced by physical and chemical processes have a lower potential for recovery than those that are produced by mechanically joining elements that can later be disassembled.

It will likely prove difficult to favour structural systems capable of disassembly in larger, more complex buildings. Impermanent structural solutions will generally work on a small scale where the choices of materials, as well as of removal and recovery processes, are made by the architect and client. In larger projects like high-rise buildings and other intensive building types, a whole host of additional factors will have to be included, including safety, stability and fire protection. The need for structural stability will also often dictate the use of those building systems that are the most difficult to disassemble. Furthermore, these large projects are designed within parameters established by various government bodies on the local, state and national levels, all of which will have to be satisfied by

The multi-component product

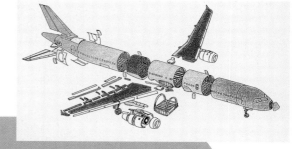

In thinking about a tall building designed for recovery, it will be useful to recall our earlier caveat concerning the completely cyclical resource system. It will likely be difficult and probably undesirable to design a skyscraper and other similarly large buildings that are fully capable of being disassembled and recycled. The structure and mechanical systems of such a building would probably be overly sophisticated and redundant. Recovering all the parts of the building would likely result in unwarranted expenditures of energy and materials.

the design. Certain means of achieving structural economies may make the production phase easier and more economical, but they will make dismantling and salvaging the building more difficult. Such processes include the use of continuous structural members, joints made in situ with physical and chemical processes and the use of complex components composed of multiple elements.

The designer must nevertheless consider the way in which the built structure will eventually be demolished of dismantled. The demolition methods will naturally have a significant impact on the qualities of the salvaged materials, as will the type or source of obsolescence. For example, a building or component made obsolete by weathering will also have been physically decayed by the elements. For components to be recycled, they must be in good condition or at least be capable of being repaired. They must also be compatible with new construction, particularly with regard to their dimensions, capability for performance and method of installation. With the eventual destination of the recovered elements in mind, the designer in effect pre-selects how to demolish or dismantle the building. He or she will want to know whether the component parts of the building are to be reused or whether the materials are to be salvaged for regeneration. The decisions made concerning dismantling are crucial, since it is at that point in the cycle of materials that wastes are transformed into new resources. Unfortunately, the vast majority of buildings now standing have not been designed with the ease of eventual dismantling and recycling in mind.

For a potentially wasted element of the built environment to become a resource, there must be a present-day need for it. Thus, a sort of design that imagines the ultimate recovery of building materials and components presupposes a continued need for those parts of the structure. If the elements of a building are not related to a material that has been used over a long period of time and that can be incorporated into a wide range of structural systems, for instance bricks, which have been used from Roman times to the present, then recovering them will be pointless. The recovery of materials of limited utility will only produce expensive waste. The geographical presence of the resource, and the existence of a local market for the salvaged elements, will also have an impact on the economics of recovery.

Finally, the designer will want to choose which kinds of operational systems to use in his or her building carefully. The mechanical and electrical systems used will have an impact on the amounts of materials and energy consumed by the building during its useful life. Cyclical systems that reduce the consumption of resources and production of wastes should be favored (see chapter 7).

To facilitate the recovery of the outputs from the built environment and their cycle of activities, the total pattern of use has to be reviewed. It could be stated that our entire manmade environment needs to be restructured so that it can become more efficient in its use of energy and material resources by recovering the bulk of the resources that are within that environment. It is generally held that a human life-support system based on a recovery pattern of use is the only structure that could operate for an indefinite period of time in a finite system. The feasibility of the recovery of any output depends on the technical means, the product output specification (or the design programme), and the availability of a potential use (or demand) for the recovered output.

The green designer's strategy should therefore be to design for recovery. This means that the design should ensure a cyclic pattern of use of materials in the building along the following lines:

● Design for reuse
The reuse of the material can be either primary (in its original form) or secondary (in modified form). Primary reuse means that the item is reused for the original intended purpose and does not require any additional reprocessing. Secondary reuse involves employing an item again but for a different purpose, and thus requires modifying it in a limited way. Secondary reuse involves creative reuse solutions for parts of the building that no longer serve their primary or intended function. Planning ways to reincarnate these components as part of the initial design phase would enhance both the probability and speed of their eventual reuse. The secondary use is predetermined and planned for. Reuse is preferred over recycling as it uses less energy and effort.

We should be aware of the danger of focussing on one aspect of use, which can give us a misleading picture of its overall performance. The design must be seen within a life-cycle overview. Generally, the method of construction and assembly of materials affects reuse and recycling. The design of the connections and fixing of materials should take into consideration the following:
● Make the components easy to disassemble (e.g., mechanical methods of fixing).
● Reduce the number of different types of materials used.
● Avoid using combinations of materials that are not mutually compatible.
● Consider how materials can be identified (in the long term, some form of chemical tracing ingredients may be used).
● Ensure that it is possible to remove easily any components which would contaminate the recycling process (e.g. microprocessors).

● Design for recycling

Recycling is a resource recovery method that involves the use of an output after it has undergone some reprocessing, accompanied by a complete or partial change in form. This can also involve the collection and processing of waste products for use as raw materials in the manufacture of raw products. This is, however, more complex than simply designing for reuse, because the collection and distribution infrastructure for recycling may incur additional energy costs. To justify the recycling of a material, the designer must ensure that the energy and resources saved (and the reduced impact on the ecosystem) are greater than those needed to make a fresh product. We should note that composite materials make recycling difficult.

● Design for durability (to increase the life of the material or the built system)

A long-life product which is easy to reuse or to repair means less overall waste. Designing products so that they last longer than their predecessors is one way to reduce waste through reuse. Sometimes this can be done by employing a new technology, as in energy-efficient compact fluorescent light bulbs, which last much longer than traditional incandescent bulbs. Another approach is to fabricate products using more durable materials.

● Design for efficient material usage

This technique simply means an efficient use of materials by designing to reduce the amount of material used (if material is scarce or non-recyclable). It does not mean a minimalist design approach, though we should seek to conserve rare materials as much as possible or design for recycling and reuse.

● Design to minimise waste

This will require a good knowledge of the life cycle of the product and good information about the performance of different materials within the reuse or recycling chain. It also raises fundamental questions about the wisdom of designing products that have a life expectancy far shorter than that of the materials of which they are made.

● Design for re-introduction into the natural environment

This means designing to ensure that the materials after use are biodegradable and can be reintegrated back into natural systems.

● Design for remanufacture

This involves the partial or total reconstitution of the output into its original form prior to use. This concept is sometimes referred to as 'products-in-service', where the supplier of the product or material will buy back the material after its use for remanufacture. The material is thus regarded as purchased in a form of 'lease' to the user. The supplier assumes responsibility for the eventual remanufacture, recycling or disposal of the material. Parts of the building can be designed at the outset to be disassembled, refurbished and reassembled when they eventually wear out, using some new parts or parts retrieved from other products. Remanufacturing is commonplace in the automotive and defense-related industries. Remanufacture for the U.S. military alone has an estimated yearly value of $7.5 billion. Remanufacture can save up to 70 percent of the resources, labour and energy used to produce and distribute new products (Goldbeck, op. cit.).

Designing for remanufacture involves:
- Ensuring that parts are interchangeable between items.
- Making components repairable or easily replaced.
- Allowing for technological components to be replaced without affecting the overall frame of the product.
- Choosing a design aesthetic that allows for the easy update of that part of the building through the replacement of a few key components such as panels.

- Design for repair and maintenance
Designing parts of the building (e.g. the cladding) so that they can be conveniently maintained and repaired entails the availability of replacement parts, as well the ability to take these parts apart and put them together again easily. Easy-to-follow manuals and accessible technical assistance from manufacturers can be instrumental here. But more than just designing to facilitate repair and maintenance, we should be aware that improper use, overuse and neglect all shorten a component's life, minimising reuse potential. It is therefore beneficial for all users to learn how to handle and operate the building and its components and systems correctly, utilise them reasonably to serve real needs, service them when called upon and store or protect them from the climate safely and intelligently.

Timely repair to the building can extend the life of its components. The opportunity to keep the building and its systems working is a vital component of reuse. Many building components are manufactured in a manner that makes them almost impossible to repair. For example, it's extremely difficult to service or to replace components that are welded together, chemically bonded or riveted (rather than fastened with screws that allow disassembly) or are permanently sealed in a housing. As a general guideline, when the cost of repair is no more than half the cost of replacement, repair is always preferable.

● Design for upgrading
A number of items in the building are amenable to upgrading as owners' needs change or technological advances occur. For example, the building's BAS (Building Automation Systems) computers can commonly be upgraded by adding a larger memory chip, a new drive or other similar parts. Several copying machines are also now being designed with the capacity to upgrade in mind.

Responsibility for the Final Sink for All Outputs

The typical large building or skyscraper generates outputs throughout its life cycle. The difference between manmade systems and natural systems is that natural systems do not produce waste. 'Wastes' in natural systems are reintegrated or assimilated into the natural cycles in the biosphere. Manmade systems, because they have not been designed for recovery at the onset, are waste generating. For example, in many production processes in the built environment, the material outputs comprise up to 25 to 50 percent by weight of the outputs (Bower, 1971). Outputs from construction processes include the rubble of the structure, concrete, bricks, timber cutoffs, metal work, etcetera. Solid outputs from the residential skyscrapers during their period of use consist mainly of paper and fermentable organic matter but also frequently dust, cinders, textiles, glass, porcelain, wood, metals and plastics. Outputs from commercial buildings during their period of use are largely paper-based, but also include food wastes. Outputs from industrial buildings include building wastes, plastic, wood textiles, ash, gaseous emissions, liquid effluents from production processes and other toxic discharges (see Fig. 47). As the activities of the manmade environment increase in volume and diversity, so will the forms of output or waste.

The crucial point to recognise is that, no matter how well our built environment has been designed and operated, there will inevitably be end products that need to be disposed of. Acknowledging this fact is vital to ecological design.

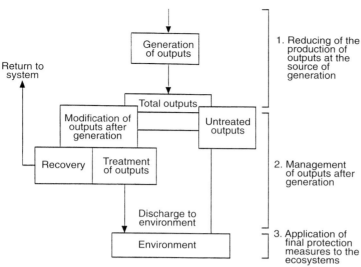

Fig. 47   Outputs from
the built environment
(Source: Yeang, 1995)

Map of the possible routes taken by the outputs from the built environment.
This is a generic model for tracing the flow of outputs from the built environment;
it could be considered as a problem definition tool. No equivalent status is accorded
to each of the measures listed. The selection of the appropriate form of management
for each individual output from a system will thus depend on the built system, the
form of outputs discharged, the operating conditions, the form of the inputs, the
state of the environment, and the interaction of all these factors.

Although the designer may be able to reduce, change or convert
the form of any output, what finally remains has to be eventually
disposed of and sent back into the environment. The designer cannot expect these unwanted outputs to disappear, nor can he or she
expect to throw them away elsewhere. Although some outputs may
be kept in use cyclically within the built environment by means of
recovery processes, it should be remembered that the final sinks for
any unwanted output are the ecosystems in the biosphere. Green
design must guarantee recovery (see previous section), but if this is
ultimately not possible despite our best endeavours, then the
designer must design for the ecologically sustainable disposal of
these outputs.

We can speak of the built environment as being engaged in
internal-to-external exchanges with the ecosystem in those cases in
which any kind of energy or materials are expelled from buildings
into the environment (i.e. L12, Wann, 1996). Such wastes discharged
by the built environment are absorbed into the ecosystems, in some
cases after having been treated. Such exchanges between the built
environment and the earth's ecosystems constitute the internal-to-
external 'transactional interdependencies' of the former.

The management of materials expelled by the built environment into the ecosystems is not ordinarily in the domain of the designer. Rather, related disciplines have dealt with such environmental protection issues as the disposal of solid wastes and pollution control, including air and water pollution engineering and the disposal of liquid effluents (e.g., Hannon, 1973). This specialisation makes it difficult to consider the emissions of the built environment in a holistic way. Despite this structural impediment to solving environmental problems, it has nevertheless been recognised that exchanges of materials and energy need to be monitored. Evaluating such emissions is the first step toward improving those situations in which the ecosystem is threatened with contamination.

The interactions framework (see chapter 3) informs us that the discharge and the management of outputs are related systemically to the flow of inputs, the operations within the built environment and the assimilative capacity of the earth's ecosystems. In the design process, it is necessary for the designer to anticipate the net outputs associated with his proposed designed system (over its complete life cycle), the related impacts and interactions with ecosystems, the extent and type of inputs used in managing these outputs and the ways they can be managed cyclically within the built environment. The total outputs of a designed system in relation to its life cycle are represented in figures 1 and 2 on pp. 7–8.

Every open system generates outputs and expels them into the environment (Bolitho, 1973). The disposal of waste products will thus concern every human society, even those whose economies are based exclusively on agriculture: even ploughing a field produces some amount of runoff of organic materials, and even simple sources of heat for cooking such as wood fires generate smoke and carbon dioxide which are released into the air. Human habitation of the planet will inevitably result in organic waste. The important environmental question then becomes the management of such waste products.

Given that human beings will inevitably produce wastes, we must try to ensure the continuing viability of our ambient ecosystems through design and planning. Living systems can only remain stable when their environments are able to absorb their outputs as they are produced. When no attempt is made to control outputs, the natural environment is disturbed by waste products and 'random' parts. If an environment comes to be constituted only by random parts – when complete entropy reigns within it – then it will no longer be conducive to life. If human beings knowingly contribute to this kind of destruction of the environment, they should know that they are contributing to their own eventual demise. Given the nature of the biosphere as a closed system, it is clear that an abundance of products from one system will necessarily displace those of another. If we fail to address the question of disposal of the materials human beings produce through their settlements, then pollutants will continue to strain the ecosystems and the environment as a whole will be further degraded.

Human beings have the potential to destroy the environment, and thereby themselves, or to reduce pollution generated by their built environments. If we solve the problem by reducing waste, then we will not have to invent new ways of disposing of it in such vast quantities. Since waste treatment systems only operate at an environmental cost, it is preferable to reduce waste production at its source in the built environment. Considerations of waste production and disposal can be a part of the design process from the outset if potential impact on the environment can be assessed in advance. In predicting what kinds of wastes will be produced by a building in the planning stage, it is useful first to classify potential pollutants. There are several possible classificatory schemes, for instance by state (solid, liquid, gaseous, particulate) or by degree of toxicity. If we chose the second approach, for instance, we would look systematically at all hydrocarbons, compounds of mercury and so on, through the ranks of toxicity.

Another way to understand the wastes produced by the built environment would be to analyse their sources. For instance, outputs might include the byproducts of the built environment's construction. Among these types of materials would be waste lumber, main tailings, rejected building elements and so on. Such materials can be collected, and if reuse is technically possible, they can be distributed to new sites. If the materials cannot be reused, then they will have to be disposed of, but this must be done carefully in order to minimise environmental harm. Some forms of energy and materials generated by the built environment are by their natures impossible to recover. It will be impossible to entirely prevent energy loss from friction, heat loss, and other forms of output. For these kinds of energy exchanges, the best we can do is try to minimise them through good design. Buildings themselves, when they reach the ends of their useful lives, can be recycled, as can building materials, materials associated with demolition, product packaging and other potential wastes. Building, maintaining, and demolishing parts of the designed structure will also generate materials, like dust, that can be collected and recycled. Given this vast array of materials and energy generated by every building activity, it is no wonder that pollution has been dealt with in such a piecemeal fashion.

The sheer number of potential pollutants has also frustrated conservation attempts.

Limiting the quantity of materials and energy expelled by our large building into the environment will become a central objective of our design method. Some distinctions can be made among the non-recyclable materials and quantities of energy that the building is inevitably going to produce. On the most basic level, some of these wastes will cause an increase in the volume of a particular material or energy source that already exists in the environment. In other (and worse) cases, potentially poisonous foreign materials will be spewed into the environment. The second category of wastes is obviously to be avoided by all possible means (Odum, E.P., 1971, p. 75).

The Management of the Outputs from the Skyscraper and the Intensive Building Type into the Environment

Much of current ecological design endeavours have focussed upon energy efficiency or conservation; greater importance need to be placed on outputs management, which is to say the 'internal-to-external exchanges' from the built environment into the natural environment (L12 in the interactions framework). As we have mentioned earlier, to design our building as a completely closed system, without any exchanges of energy and materials with its external environment, is not possible in practice, since external environmental interactions are necessary attributes of living systems.

Having exhausted all options for designing for recovery (see above), we may have to accept that some amount of environmental interactions and exchanges cannot be avoided, and we would need to ensure that our design process must be geared towards the reduction of initial energy outputs. This focus is crucial since, once an output is produced, it can only be managed at the cost of additional energy, materials and impact on the environment. In the case of the skyscraper and similarly large buildings, it is important to work at the outset to bring about a reduction in the amount of materials and energy that will be expelled by the building. Once the building is completed, certain limitations (discussed above) will impinge upon the recovery process.

The designer is in fact the 'outputs manager' for any given building project, who works to minimise its negative environmental impact. First, the designer has to determine just what the part of the built environment under his or her control is going to expel into the surrounding ecosystem. Using the interactions framework (see chapter 3), the designer considers the whole range of emissions from the building. The designer has to know not only what materials and energies are going to be produced by the building(s), but also what forms they are going to take, with what parts of the natural environment they are likely to interact, what their effects on existing life forms are likely to be, and finally, where they will ultimately end up. The decisions a designer makes concerning these potential pollutants are fundamental and must be integrated into the design process from the outset. As an outputs manager, the designer works throughout the process to control the amount of materials and energy expelled, modifying the design in order to reduce pollution. In those unfortunately unavoidable cases in which pollution from the new project cannot be helped, the designer must take steps to ensure treatment of outputs.

Decisions that affect pollution by the built environment are not presently considered to be the designer's prerogative in most instances. Instead, the question of emissions is ordinarily shifted to others, such as pollution engineers, who sometimes rely too much on simplistic technological solutions to the problems that do not take account of the ecological complexities of our ecosystems.

An ecologically simplistic approach limits the kind of broad consideration of environmental consequences described above. If, for instance, any activity threatens to produce water pollution, a typical response would be to apply technical solutions to the problem. These might take the form of diluting the anticipated water pollution or treating the wastes prior to their emission from that activity. The ecological approach, on the other hand, would instead alter the activity at the source in order to reduce or eliminate the amount of pollution produced in the first place. Unfortunately, this second approach is rarely taken.

If the designer is operating with regard for the ecological consequences of his or her work, he or she will evaluate all of the projected emissions from the building and their impact on the environment. Of course, it may not be possible to anticipate all of the ramifications of operating the building. Nevertheless, the obligation to evaluate the quantities and consequences of all of the materials that will be discharged from the building remains.

Methods for managing the outputs cannot be considered before this is done.

As the designer attempts to gauge the impact of the building on the natural environment from the earliest stages of the design process, several variables should be taken into consideration. First, the designer must know what is going to be discharged. He or she must determine the sources of outputs as well as their qualities and quantities. Second, the designer will want to know where the emissions will have an impact. Third, the designer must ask what type of damage will be caused and how regularly. Fourth, when the designer has determined the extent and character of the impact, he or she will have to assess whether it is significant. Having resolved that the impact is significant, the designer will then analyse the range of solutions to the problem and propose various design responses aimed at limiting the pollution. Finally, once the building is in operation, the designer will want to confirm that the measures proposed for limiting emissions have been adopted and will also monitor their effectiveness.

We should therefore regard the management of outputs from our building as a problem to be addressed ideally within the building itself, or at least in the immediate context. In the case of an urban high-rise, that will mean dealing with pollutants within the city itself. The transfer of emissions from the urban environment to outlying areas should only be contemplated in cases where the pollutants cannot be reused locally or where they cannot be absorbed by the local ecological environment except at the cost of contamination and excessive expenditures of more energy.

Our partitioned matrix (in chapter 3) clearly suggests that our design response to the reuse or recycling of potential waste products should be determined by balancing three objectives. These are: minimising the amounts of energy and materials imported by the building, decreasing the potential for wastes to be generated by the building and minimising adverse effects on the ecological environment. In balancing these objectives, we should attempt to minimise the generation of wastes rather than focus on the ways in which they can be contained or treated. These latter processes are less effective ways of reducing strain on the earth, since they will necessarily require additional energy expenditures. It is best if the designer instead makes choices that lead to pollutants being minimised or entirely eliminated so that methods do not have to be invented or systems put in place to take care of them.

In those cases in which the designer cannot entirely preclude waste outputs from his or her building project, specialists in recycling as well as pollution control engineers can use indicators to assess pollution levels. They can then determine the capacity of a given ecosystem to assimilate permissible forms of pollution. For example, for water pollution, algae blooms can indicate the presence of dissolved oxygen, evaporation, nutrients, and fecal coliforms, as well as pesticides, herbicides and defoliants. The pH of water can also be analysed, as can the physical characteristics of a given body of water – such as sediment load, stream flow, temperature, turbidity, and so on – to reveal the presence of toxic and non-toxic dissolved solids.

In addition to water, pollution engineers can also analyse the air and earth to determine contaminant levels. Indicators in the atmosphere can reveal the presence of various forms of air pollution, including carbon monoxide, hydrocarbons, particulate matter, photochemical oxidants and sulphur oxide. The use and misuse of the land itself, which we might term 'land pollution', is indicated by obvious forms of soil erosion and soil pollution.

Although it is convenient to think of indicators in terms of the parts of the ecosystem that are open to damage – earth, air, and water – the impact of pollutants on the entire system is also significant. The effect that various discharges from the built environment have on species of plant and animal life, populations, natural habitats and communities, as well as on the larger functioning of the ecosystem, should also be considered.

The pollutants that the pollution engineer finds in the environment are in fact a consequence of the decisions made in the earliest stages in the design process. From the moment that the designer makes a first schematic response to the requirements of a particular programme, he or she is determining what environmental impact the building will have. From the beginning, a built system should be designed in ways that minimise the production of those materials and quantities of energy that will contribute to the destruction of the environment. If this is done, then the building will not further tax the capacity of the ecosystem to absorb outputs (see chapter 2). Since the environments that surround most of our intensive building types such as skyscrapers are densely built, and their absorptive capacities are strained already in many instances, it is essential that new high-rises and other urban buildings not produce large quantities of energy and materials that have to be assimilated.

Many of the common building materials are associated with the industrial processes having the most serious potential for environmental pollution. One argument for their continued use is that many of these materials are potentially very durable, and if they are reused, the disadvantages of their manufacture can be balanced by their longer useful lives.

The appropriate measures for output management can be conceived in the following ways:

- reduction of outputs at the source of generation
- management of outputs after generation
- application of final protective measures (i.e. reintegration)

These measures may be related to the pathways taken by the outputs, as shown in figure 48.

To reduce the amount of waste materials produced by the built environment at the point of their potential generation, the designer can either modify the design or alter the methods used to manage materials expelled from the building. The first step is to reduce amount of materials and energy produced by the building and to ensure that those that are expelled exist in forms that are susceptible to effective management. Achieving these goals will mean modifying the project's design, and carefully selecting building materials and methods. At this early stage the designer takes steps to ensure that everything generated by the building can be managed effectively.

There are several means of reducing the amount of waste products from a building. First, the designer considers what materials and energy sources are flowing into the building. These inputs should suit the anticipated patterns of use of the built environment as well as its projected life span (i.e. L21 in the partitioned matrix in chapter 3). This fit will be easier to achieve with projects that are in the design phase, but even for existing buildings, the designer can take measures to ensure that the materials and energy used produce relatively little in the way of waste products. Second, the designer can modify the building programme in order to reduce the production of outputs. Third, it may be possible to reduce levels of energy consumption by the building's users, which will in turn lead to an overall decrease in

Fig. 48   Pollution control (Source: Yeang, 1995)

In pollution control, for example, certain indicators of the levels of permissible pollutants are often used to determine the assimilative capacity of the ecosystem. Such indicators, for instance, include:	
Form of pollution	Indicators
Water pollution	Algae blooms
	Dissolved oxygen
	Evaporation
	Fecal coliforms
	Nutrients
	Pesticides, herbicides, defoliants, pH
	Physical water characteristics
	Sediment load
	Stream flow
	Temperature
	Total dissolved solids
	Toxic dissolved solids
	Turbidity
Air pollution	Carbon monoxide
	Hydrocarbons
	Particulate matter
	Photochemical oxidants
	Sulphur oxide
Land pollution	Land use and misuse
	Soil erosion
	Soil pollution

by-products being turned out by the building. Fourth, the processes taking place within the building (L11) that are responsible for
outputs can be changed. For instance, the construction, production
and recovery processes could be modified to increase efficiency and
thereby reduce waste. By using less energy and materials in the
built environment as a result of increased efficiency, we limit the
destruction of the ecological environment. Fifth, the designer can
adjust the durability of the building to suit its anticipated use and
life span. Finally, the internal processes of the built environment
(L11) should be coordinated with the ability of the surrounding ecosystem to assimilate the projected levels of emissions. Not only will
the level of outputs need to be considered, but the timing and regu-
larity of emissions will have to be taken into consideration as well.

The management of waste products in the built environment
usually requires additional expenditures of energy and materials
with some added environmental cost. The designer should therefore ensure in the design and planning of the built environment
that difficult outputs should be minimised or should not be permitted to be generated in the first place, as their disposal would
inevitably have to be considered (see Fig. 49).

For a building that does expel materials or wasted energy, the
designer has the option of dealing with this pollution either before
or after it is actually discharged into the surroundings (see Fig. 50).
Some methods for managing outputs after their generation are discussed below. In some cases, it may be possible to devise a use for
materials or quantities of energy that might otherwise be expelled
into the environment. In such cases, these potential outputs will be
reused, recycled or remanufactured. (For further clarification, see
'Selection of Materials' above). Where the design for recovery (e.g.,
reuse, recycling, remanufacture) cannot be adopted, the final
options consist of protective measures.
These will include the pretreatment of pollutants – those by-products that are impossible to reuse in any way. By considering
such measures, the designer will be minimising adverse impact on the environment.

One area of savings is in containers.
Those parts of the building that involve
containers can be designed for refilling.
Refilling eliminates the need for making
(and disposing) of replacement products or
packaging. There are three basic refill systems: (a) the container can be returned to

Fig. 49   Estimated total
annual waste (UK, 1980)
(Source: Barton and
Bruder, 1995)

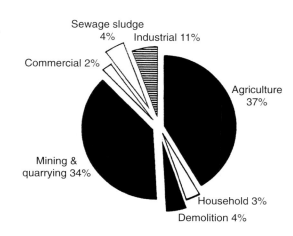

**Fig. 50   Closing the loop
(Source: Yeang)**

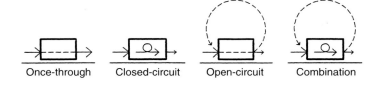

the manufacturer for refilling; (b) the user can take the container to the refilling point; or (c) the user can buy refill packs, which contain less packaging, to refill original containers. Many of the supplies that go to a building (e.g. food and fuel) come in containers.

The design will also want to consider the issue of replacement. Some parts of the building may be made of durable materials but fail to endure simply because one part wears out before the rest and can't be replaced. The capacity to replace these crucial parts can be incorporated into the design. Many common building products could be made reusable through basic design changes. One way is by modular construction, which can make repairs quicker and easier to accomplish: when one component fails or becomes obsolete, the particular module can be easily removed or replaced. With modular construction, the users themselves may even be able to do more of their own repairs and replacements. In the case of professional servicing where labor is often the primary expense, modular assembly can mean lower repair bills. A related concept is design for disassembly (DFD), which is aimed at making it easy to take apart complex products and components.

The designer must be aware that to design for recovery, extra storage space must be provided in the building. At the minimum, there must be adequate storage space for the building-wide collection of separated glass, aluminium and office paper.

In those cases where it is impossible to recycle potential outputs, the designer can also consider their pretreatment. Through pretreatment, the harmful qualities of emissions are neutralised. The material or energy is modified so that it causes less strain on the ability of the ecological environment as it is absorbed. Pretreatment methods can comprise either physical, chemical or biological transformations (Tebbutt, 1971). Physical processes improve the materiality of the expelled material; for example, the size(s) of the particles that are discharged, the specific gravity of the materials, their viscosity and so on. Discharged by-products can also be made less harmful by altering them chemically. Such transformations can be brought about by reagents, which also must be chosen carefully since they can create additional environmental harm. Finally, biological processes work to isolate colloidal organic impurities and other materials in solution in a material that is expelled from the built environment. Biological reactions can also be staged in order to make the material more susceptible to absorption into the environment. The processes of filtration and sludge activation are both biological in nature.

Regardless of whether the process used is a physical, chemical or biological one, pretreatment nearly always requires additional energy and materials. Pretreatment also runs the risk of further environmental consequences. Our objective, therefore, in designing a pretreatment system is to ensure that it can function with minimal harm to the environment. The process should also not completely exhaust the assimilative capacity of the environmental sink into which the products are going to be discharged. Pretreatment is not a panacea: it does not reduce the total output of the built system. Rather, pretreatment results in materials that are better suited – by virtue of their forms and the temporal and spatial patterns of their discharge – to be assimilated by the environment. Ideally, pretreatment should turn pollution into benign discharges, but it should only be contemplated after all possible steps have been taken to reduce the amounts of resources used and expelled by the building, and after a thorough analysis of the surrounding environment has been made.

Some potential emissions (for instance, hazardous material) cannot be treated, and therefore cannot be readily disposed of. For systems that are going to produce such materials, it is perhaps best to design a retention system. The offending outputs can then be stored temporarily until methods and opportunities for disposal are more favorable (NAS-NRC, 1966). This kind of design can also be employed to attune discharges to the assimilative capacities of the environment: materials can be stored until the receiving environmental sink is able to handle them. Hoppers, tanks and other storage systems can also be used to separate pollutants while awaiting further processing (Chappel, 1973; Bolitho, op. cit.). In other cases, by storing wastes we can reduce their volume while we await better solutions for the removal, treatment, recovery or discharge of the materials. Obviously, storage will require space and equipment. It is best not to have processes that produced these outputs in the first instance.

Storing emitted materials is only a partial solution, since it does not fundamentally alter them. Unfortunately, the most environmentally problematic materials – toxic wastes, hazardous and radioactive materials – have been dealt with in this manner. In addition to being stored in a variety of containers, these materials are sometimes subjected to compaction or refrigeration. Storage and treatment can take place at transfer stations or processing centers, or on-site.

A nearly opposite approach to pollutants from storage is the dispersion of them over a wide area. In certain circumstances, the designer may opt for casting the discharges over a large expanse of land, water or air. The concentration of the pollutant is thus lowered to the extent that it has a negligible adverse impact on any one area (in Koenig et al., 1971). For instance, to disperse gaseous discharges from an industrial plant over the widest possible area, a designer might recommend the construction of a high chimney. Or the industry might be spread over a larger area rather than concentrated in one place, in order to disperse the pollutants to a greater degree. The siting of industrial plants can also be determined with particularly sensitive areas in mind (Bower et al., 1968; Holdgate, 1972). Ideally, dispersion will not be counted on alone to solve pollution problems, since it does not essentially reduce the amount of harmful discharges released into the environment. As we are dispersing a pollutant, we should also be looking into ways of reducing its volume or eliminating it by redesigning the built environment.

Environmentally harmful emissions can also be diluted in order to minimise their impact. Increasing the volume of some substances through dilution in a benign medium can sometimes make it easier for the environmental sink to accept them. Furthermore, the many sections of the environment differ in the degree to which they are capable of accepting pollutants. It may therefore be helpful in some cases to divert outputs from their places of generation. Thus the materials can be transported from an area that is unsuitable for discharge to one that is capable of assimilating the discharge, possibly as a result of its lesser degree of previous contamination. This alternative should not be used excessively, since the environment is a finite area whose supply of unpolluted portions will eventually be exhausted. Furthermore, when pollutants are transported they consume even more energy and cause additional environmental harm.

When all preventive measures have been rejected as impracticable, the designer has no alternative to 'final protective measures'. With the knowledge that potentially harmful substances are going to be generated by the built environment, the designer takes steps to protect humans, animals, plants and other inhabitants of the natural environment from harm. Designing these protective measures can take two forms: designing for environmental treatment and designing for environmental desensitisation.

Designing for environmental treatment consists of making changes to the receiving sink such that the impact of discharges will be minimised. This is done in place of treating the outputs at their points of generation, and it is particularly useful as a means of counteracting the cumulative effects of a variety of emissions which will all end up in the same sink. The designer should seek as much as possible to anticipate the consequences of disposing of pollutants in the environment through a careful analysis of impact on the local ecosystem. There are obvious drawbacks to this approach: it will require additional energy and materials, and it will offer only a limited short-term solution to environmental pollution problems. These drawbacks are of relatively less concern when responding to singular disasters than to long-term environmental issues. For instance, this approach could be usefully employed in the case of an oil spill when the environment in the area of this one-time accidental discharge could be modified to help it assimi-late a greater amount of oil (in Spofford, 1971).

Designing for environmental desensitisation entails making a receiving sink less sensitive to the impact of contaminants. This sort of design attempts to protect those elements of the environment that are susceptible to harm – including human beings, flora, and fauna – from their degraded surroundings. In some cases, these potential receivers of the negative impact can be separated from the source by a buffer zone in the environment. Another example of densisitisation would be spraying a scent on a contaminated beach to lessen odours from pollutants. As is clear from this example, this approach is a last resort for environments that are not susceptible to any other methods, and it will certainly require additional energy and material expenditures.

Reintegration of Outputs into the Ecosystem

Where it is unavoidable and discharged outputs enter the environ-ment, they can be treated in one of three ways: by being transport-ed within the ecosystem, by transformation from one form to another or by storage (after Bower and Spofford, 1970). Each of these approaches will have an impact on the environment. Trans-portation will require the use of an environmental medium, trans-formation will demand energy to accomplish the modification of the material's physical, chemical or biological makeup, and storage will take place in some part of the environment which may suffer an adverse effect.

When a building discharges a given substance, that material (or amount of energy) can take one of a number of routes through the natural environment. These potential pathways must be analysed before the discharges are made, because even the most minute and seemingly inconsequential amounts of outputs can have serious environmental implications. For example, a 1.25 °C increase in the surface temperature of an aquatic system, brought about as a consequence of thermal emissions, could have a serious impact on biotic communities living in the water (Waggoner, 1966). When the impairment of the ecosystems by a discharged material becomes significant, an ecological imbalance has occurred. Thus pollution is constituted by the entry into the ecosystems of materials or quantities of energy, produced by human beings, which harm or destroy those ecosystems. Pollutants can be substances that are not native to the ecosystem, or they can be materials or forms of energy that are familiar in a given ecosystem, but not in the large concentrations that result from human habitation and which are unhealthy for the environment (cf. Ashby, 1971).

Obviously, the degree of environmental impairment from a given pollutant is related to the ability of an ecosystem to assimilate that discharged material or energy. The assimilative capacity of the ecosystem with respect to any given pollutant will be variable over time and place. It will depend on local conditions as well as on the stochastic qualities of certain elements of the ecosystem. For instance, stream flow, temperature, amount of light and other factors will influence assimilative abilities; the character of the discharged materials or energies will also play an important role in determining the ability of the ecosystem to absorb such outputs.

Whenever the amount of harmful energy and noxious materials discharged into the environment exceeds its ability to absorb them, this increase in pollution sets off a chain of adverse environmental consequences.

As we have already observed, a great amount of damage to one part of the environment may have a ripple effect on another. A pollutant can move throughout the environment, leading to changes in the qualities and quantities of resources in various areas. When a pollutant arrives in an ecosystem, it may upset the existing equilibrium, which then must be re-established. In cases of minor pollution, the disequilibrium may only be temporary: the pollution may cease and the system's equilibrium will be reinstated. But if a serious amount of pollution appears in an ecosystem and is allowed to continue, and the output exceeds the assimilative capacities of the system, then severe environmental damage can occur. There may even be destruction of the ecological environment, but in any event it will be damaged. In the latter case, the ecosystem will only contain those organisms that are able to survive in these diminished circumstances.

The impact of discharges on the ecological environment is not an issue we can moot through sensitive design for recovery. Not all materials and quantities of energy can be reused within the built environment without exerting even greater economic and environmental costs. Therefore, designers will have to determine the levels at which outputs can be discharged into ecosystems without overtaxing their assimilative capacities and without harming them in other ways. Most effluents may well have to be expelled into the environment; it behooves us then to assess at what levels this discharge can take place safely.

In order to assess the assimilative capacity of the environment in the location where the discharge is to take place, the designer must determine the biological and physical capacities of the local ecosystem. He or she must also understand the characteristics of the materials to be expelled, as well as the patterns of their discharge (see chapter 5). Thus the designer has to be able to specify the times and places of discharges in order to map concentrations of potential pollutants in the environment. Making such assessments is difficult, since emissions do not come from a single source but from a variety of them.

In assessing an overall pattern of discharges a number of factors will have to be played one off the other. An extreme example would be provided by an area in which heat was provided overwhelmingly by electricity, in which transportation was largely by electric-powered vehicles, in which gases from industry and steam plants were wet-scrubbed, but in which garbage was ground up and then dumped into the sewerage system and disgorged in raw form into waterways. In such a place, the air would be protected to a large extent. However, this area would be taxing its aquatic environment to the extreme. To take a contrasting example, a region in which municipal and industrial waste-water was treated effectively, and in which sludge and solids were incinerated would be protecting its aquatic environments; however, it would be straining the assimilative capacities of the land and air. A third example would be a region that encouraged a great deal of recovery and recycling of waste materials while also achieving a low level in the production of residuals. If this degree of efficiency characterised the entire area and all production processes, there would be very little in the way of discharged pollutants entering the environment. Yet this region would require large quantities of materials and energy to fuel this level of recovery and recycling. The interactions framework in chapter 3 helps us to make the difficult trade-offs that are required to achieve a balance in the area of environmental impact.

However, at the end of the intended or designed economic life, the ecological physical life of our building's components, material or equipment persists; therefore, some form of reuse or recycling must be identified for this final 'waste' to be managed. In order to avoid excessive discharges of wastes into the ecosystems at the end of the building's economic life, the extended use of the building element within the built environment is crucial.

The ecological approach considers synoptically the use of materials and energy by the built environment and its users, and the route that every material and component in the designed system takes must be managed and monitored not only in its economic context but in its ecological context in terms of its physical life (from source until assimilation back into the ecosystem).

As described earlier, the design and creation of a building becomes a form of energy and materials management extended over the built system's entire life cycle. This is because all building activities involve the utilisation, redistribution, and concentration of some component of the earth's energy and material resources from usually distant locations into specific areas, changing the ecology of that part of the biosphere as well as adding to the composition of the ecosystem. Therefore, green design of the skyscraper and other intensive buildings involves the identification of the ways in which energy and materials are used in the building and other similarly large intensive buildings (structure, cladding systems, internal partitioning systems, fittings, equipment and so on), and anticipate their flows throughout the life of the building up to the point of their recycling, reuse or reintegration into the natural environment (not forgetting the attendant impacts on the environment from their production in the period leading to the creation of the building).

Take, for example, the production of one bag of cement. First, land is lost in the extraction of clay and chalk; then large amounts of energy are used in the burning of materials to make that bag of cement (comparable to enough fossil-fuel energy to drive a domestic car for 30 kilometres). Further energy and pollution are created in transportation to the wholesaler and then to the retailer, and later to the construction site. More energy is used (and carbon dioxide emitted) in concrete making equipment on the building site and in the crane that hoists the cement up the building-frame. The complexities hidden within this deceptively simple example show the deeper level of consideration required of the designer who operates with ecological principles in mind. Similar chains of resource and energy use could be developed for every screw, nail and piece of wood or metal that goes into a built environment.

In summary, the design measures for the management of outputs can be conceived of in several ways. First, we can think about reducing the amount of outputs that are produced in the first place; this amounts to eliminating waste. Second, the designer can work to manage unavoidable emissions after they are generated. Methods of doing this will include remanufacturing and recycling. Third, we can act to protect those areas where materials and/or quantities of energy are going to be discharged.

In the management of materials and energy in the built environment, we can identify four possible design strategies of the pattern of use of energy and materials in the built-form and its servicing systems. These are: once-through design; design of open circuits; design of closed circuits; and combined open-circuit design (see Fig. 50).

As the existing built environment has been configured, it constitutes a once-through system. That is to say, resources are consumed under the assumption that they are unlimited; hence outputs are discharged with little concern for how they will effect the environment and with slight analysis of the routes they will take to their ultimate sinks.

An alternative approach is provided by the open-circuit system. Here, the designer capitalises on the ability of the environment to receive waste products from the built environment. The similarity between this system and the once-through system is that both use the environment as a receiving sink. However, in the open-circuit system the emissions do not exceed the ability of the ecosystem to absorb them. Thus, the level of discharge is held beneath the threshold at which environmental harm occurs. Presently, open-circuit design is achieved through careful geographic placement of discharges and through their pretreatment (in Holdgate, 1972). An example of an open-circuit system would concern the disposal of discharges from an industry located in a relatively pristine region with little pollution or contamination. The usual way of proceeding would be to dump the industrial discharge without regard for the land, which is considered to be an unlimited environmental sink. In an open-circuit approach, however, the environment would be evaluated and possibly modified before a system for the emission of industrial by-products would be determined.

A closed-circuit system is one in which most of the processes taking place are internalised within the built environment itself. The advantage of such a closed system is that impact on the surrounding ecosystems is minimal. Complete internalisation is possible only with respect to some outputs, since any system will rely on interaction with the environment for long-term survival, particularly for energy to keep the system operating. It also seems unlikely that discharges from the system could be entirely eliminated. Nonetheless, it may be useful to combine a closed-circuit system with open-circuit systems, especially when the assimilative capacities and other characteristics of the local ecosystems constrain a building's operation.

Ecological design should favor processes that are internalised within the human-made environment to the greatest degree possible. Yet internalisation should not be pursued to the point that it creates new environmental problems for the surrounding ecosystem. A built environment that combines aspects of the open and closed-circuit approaches has the advantage of reducing the level of environmental impact that would result from a once-through system, while simultaneously capitalising on the limited abilities of the ecological environment to assimilate discharges. As the designer embarks on the development of a skyscraper or similarly large building, he or she should be thinking in terms of these three systems. The designer should work above all else to eliminate the predominant once-through system and work toward a closed-circuit system (i.e. designing for recovery – see below). As we have observed, a totally self-enclosed system will be unlikely or even impossible, but the designer can aim toward that objective, for instance by designing for reuse and recycling. Elements of the composite system combining open and closed aspects are also likely to be necessary.

This chapter has provided the designer with a set of strategies towards the selection and use of energy and materials that would assist him or her in the design of the skyscraper and the intensive building type. Essentially, these have to follow a cyclic pattern of use where possible and with minimal energy inputs and environmental degradation, and from source to sink.

The broad design strategies in the use of materials are as follows:
- design for reuse
- design for recycling
- design for durability
- design to reduce the amount of material used (if scarce material or if non-recyclable)
- design to minimise waste

- design for reintroduction into the natural environment
- design for remanufacture
- design for repair and maintenance
- design for upgrading
- design for refilling (instead of replacement)
- design for replacement

The broad strategies in the selection of materials are:
- Potential for reuse and recycling
- Embodied ecological impact
- Embodied energy impact
- Toxicity

A design approach based on the conservation of resources can be considered in terms of three alternative strategies: (1) strategies that reduce the supply to the system; (2) strategies that impose efficiency and performance on existing systems; and (3) general strategies for the redesign of existing systems or the design of new systems. These strategies, elaborated below, are useful in looking at the designed system and evaluating its systems and materials use.

Strategies that reduce the supply and flow rate are those that:
- Reduce the existing consumption level by controlling the rate of resource use or lowering the standard of living
- Reduce flow by reducing the total number of products or components turned out
- Substitute other resources (e.g., renewable resources)

Strategies that improve the efficiency and performance of existing systems
- Encourage the recovery (reuse, recycling or regeneration) of existing components, provided it does not increase environmental degradation and contamination
- Increase the efficiency of recovery processes
- Extend the useful life of a unit or a component
- Control losses through corrosion and wear
- Increase efficiency or production processes
- Increase the efficiency of component or equipment use

General design strategies for the redesign of existing systems (or the design of new systems) aim to:
- Achieve materials and energy economy and low ecological impact by appropriate design and selection

● Redesign existing systems toward maximum performance
● Design for ease of repair and recovery (e.g., by standardisation and simplification of materials and components)
● Design for optimum life of a unit of the component
● Design for minimum use of materials per unit per component
● Design for efficiency and low ecological impact in processing and recovery
● Design for efficiency and low ecological impact in use

The above measures are of course not exhaustive of all the possible strategies of designing for the conservation of resources. They are, however, indicative of the way in which our design effort might be directed. We should also be aware of the extensive outputs of waste during construction, and the designer must ensure that the builders have a waste management programme that includes recycling and reuse of materials and responsible waste disposal.

When a designer is attuned to the ecological consequences of the built environment, he or she is necessarily concerned with the life of the building after it has been handed over to the owner or user. If the designer takes the environmental approach very seriously, he or she will be able not only to inform the owner of the environmental costs of a building, but also of the costs of its use and ultimate disposal once its useful life is over.

Operating with this seriousness of purpose, the designer will have a number of factors to consider. These will include the internal environmental interactions that result from the life of the built system. However, the impact of the built environment on the ecological one is not just a result of producing a building or complex, but also of using it. Moreover, the environmental consequences of the eventual disposal and recovery of the building must also be assessed, based on an understanding of the economics of real estate investment and accepted patterns of use. As we have already observed, the actual physical life of a building can be much longer than its economic life. The environmentally sensitive designer will be concerned with both aspects.

Patterns of resource use will also concern the designer, who must work to reduce the occurrences of linear patterns of use and will favor cyclical patterns instead. The goal of the green design process is a building or complex that minimises both its consumption of resources and production of wastes. Meeting this goal will likely mean recycling and recovery, and the designer must ensure that these processes do not require inordinate amounts of materials, energy and space in the environment.

The stability of the natural environment is a major concern for the designer who introduces a building, which is a dynamic system, into it. Once the building is dropped into the ecosystem a whole chain of interactions is set in motion and continues until the building is removed, and even beyond that point. The designer must think of all of these interactions. From the moment that building materials are procured in the environment, through the moment at which scrap materials are dumped into environmental sinks, materials and energy are consumed by the built system at the same time that discharges are made. Therefore, for a designer to operate with a complete environmental consciousness, he or she must view energy and materials management in the most comprehensive sense. The designer must understand all of the ways in which the built system will interact with the built environment over the course of its entire 'life'. Then, of course, the designer must work to minimise the negative interactions to every possible extent.

This ambitious program will not likely result in buildings in which no element is ever wasted; rather, systems should be made as efficient or benign as possible with respect to the consumption of energy and materials, and as minimally disruptive of the ecological environment as possible. This means minimising spatial impact on and pollution of the environment at the design and construction phases, as well as implementing recycling and recovery programmes. Designing with the recovery of materials as a goal may lead to a building's greater up-front cost, but in the long run, proceeds from recycling and recovery may compensate the higher expenses to some degree.

To fully appreciate the environmental consequences of building, the designer must conceive of resource use as a cyclical pattern in which energy and materials flow through the environments (built and natural). The cycle can be viewed as it unfolds from the point of production, through the period of use to the moment of recovery of the materials again. The life-cycle model enables us to relate the activities that take place within and around the building to one another. All design should be informed by knowledge of this interconnectedness: thus, the ecological consequences of discrete activities can be assessed as part of a larger system.

While it is generally agreed that our cities, buildings and other elements (such as transportation systems and infrastructure) must be interpreted as a complex system of material and energy flows, what is asserted here is that these flows must be managed to ensure that they are ecologically responsive. While the above discussion deals with physical aspects of inputs that usually concern designers, we must acknowledge that in a truly comprehensive approach we would need to take into account the inputs of food to the occupants of the buildings. On the analogy of the framework presented for creating a skyscraper and other large intensive building types, we would look at the urban food system through all of its various stages as well: agriculture and horticulture, transportation, processing, packaging, refrigeration, storage, wholesaling, retailing, display, collection by the purchaser, further processing and disposal of all wastes from the food and its packaging. Collectively, these represent a major component of total environmental impact and energy use. These aspects, though important, lie outside the scope of this work; however, they demonstrate that the ecological approach described here can be extended to all human activities and functions that take place in the built environment, and that no human activity exists in a vacuum or without an ecological context.

Chapter Seven

Designing the Skyscraper's Operational Systems

Operational Functions within the Designed System

In natural systems, the environmental conditions of the habitats of animals and other organisms are sustained by passive means (for example in anthills) or by some organic means. In the case of the human built environment, the internal environmental conditions are mainly sustained by mechanical means (e.g., using M&E equipment and technology) and very often using sources of energy that are not ambient to the project site (and thus are non-passive); instead, they are usually non-renewable sources such as electricity generated from oil, coal, gas or other fuel (see Fig. 51). Ecological design, in imitation of natural systems, tries to optimise the use of all passive systems of operation and of the climate and diurnal conditions of the locality. The operational systems in our building are

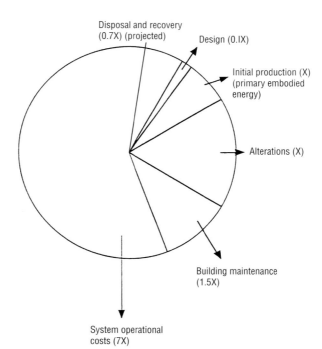

Fig. 51   Life-cycle energy costs. This illustrates the likely energy costs of a building over its life cycle, including the recovery and disposal of material and equipment (Source: Yeang)

those systems that generate internal environmental conditions of comfort and facilitate materials recovery (i.e. L11).

At the same time that the designer is selecting the materials and components that are to be used in the skyscraper and other large building types (chapter 6), consideration needs to be given to the technical means available for the operational systems. Solutions should be chosen that meet the criteria of an overall energy balance and materials recovery and reflect the latest technical knowledge on the use of environmentally compatible forms of energy (see Fig. 52). As mentioned earlier, a number of the technical solutions presented here have aspects that require refinements and modifications to meet ecological objectives fully; nevertheless, as the technology for ecological design is still in its early developmental stages, such systems are not to be seen as panaceas. By applying our 'law of ecological design' (in chapter 3), all these operational systems need to be evaluated for their inputs, outputs, operational consequences and external consequences. Many of these analyses have still to be carried out. This section provides the broad strategy for the design of the large building's operational systems. Others (e.g. Daniels, 1995, et al.) have covered the engineering aspects of various operational systems in great detail. The type of operational system to adopt (whether passive, mixed-mode, full-mode, etc.) is dependent on the climate of the locality of our building and its ecology (i.e. L22). The

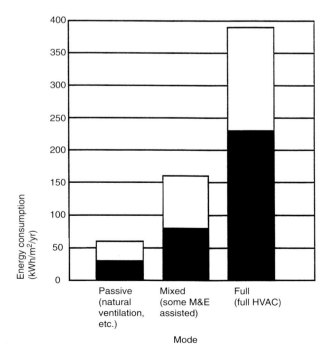

Fig. 52 Energy consumption targets of various modes of operational systems (Source: Yeang)

intention here is not to provide design guidelines for all climatic zones (which is better covered elsewhere), but simply to show how various modes of handling the building's operational systems are to be dealt with in ecological design. To illustrate the ideas, the temperate climatic zone is contrasted here with the equatorial zone.

At the schematic design stage of our green skyscraper or intensive building, we should have completed all the usual built-form design analyses for the project to enable the next set of decisions to be made, on the building's passive and active operational systems. There are the internal ecological interdependencies of the building to consider. The building's total operation activity during its life cycle must be kept in mind, and design consideration should include all the ecological impacts and interactions that result from these activities (i.e., L_{11} in the partitioned matrix of chapter 3). The important factors include spatial displacement of the environment, energy and materials (both inputs and emissions), the activities of building users and the functional systems of the structure itself. Simply stated, the total set of interactions caused by the building is not defined only by those inherent in the making and building of the building and its components (in chapter 5), but includes everything that arises from its operational use, its final disposal and the recovery phase.

As was discussed in the last chapter, we must look at the flow of energy and materials from source to sink – from extraction from the earth to disposal at the end of the building's useful life. Our examination of the entirety of the resource flow and the use of materials and energy in the built environment's operational phases should make clear each building's individual use pattern; this pattern in turn is evaluated for its potential environmental impact and modified accordingly. For example, we can establish annual energy consumption targets per square metre for different building types as benchmarks. Similar sets of design criteria can be established for the primary embodied energy or the embodied carbon dioxide production for different building types. Similar discharge (output) indexes can be established for different building types (e.g., gross tonnage of waste paper per annum per square metre, etc.).

General Strategies in the Design of Operational Systems

We have seen that the majority of energy use (i.e. inputs or L_{21}) and a similar proportion of carbon dioxide emissions (outputs, or L_{12}) are associated with buildings, with 60 percent attributed to the residential building type and 7 percent to the commercial office building. One might ask whether our design efforts should not first be directed to minimisation of the energy use in the residential building type, followed by the commercial office building type; however, these figures favour the residential building type because of the greater volume of residential buildings. On a per-square-metre basis, commercial skyscrapers consume greater energy than residential structures. In both types, about 60 percent of the energy use during the operational phase (L11) is in space heating and air-conditioning. For example, primary embodied energy for buildings in Europe (see chapter 5) does not vary much between different structural solutions (about 2.5 to 3.5 GJ/m²) and this represents only some 5 to 10 percent of the total energy over a normal 60-year building life. In effect, the entirety of constructional components (including lifetime additions) comprises only 35 percent of total energy demand. Our design objective should be to choose an architectural form and operational systems that will make the most of available natural lighting and ventilation to minimise the remaining energy demands of the skyscraper and other large building types (up to 65 percent).

The significance of the success of a green or other large intensive building type's operational systems lies in achieving a lower operational energy consumption level per annum (in kWh/m² per year). A general target for full-mode buildings (in temperate and equatorial zones) should be less than 150 to 250 kWh/m² per year for the typical conventional HVAC buildings. Added to this would of course be the necessity to meet the ecological criteria of the other sets of interactions (i.e. L22, L12 and L21 in chapter 3).

If we look at the energy consumption pie in the operations of a commercial skyscraper (Fig. 53), by far the highest energy used in its operations is in HVAC systems, followed by artificial lighting systems. The other elements (such as elevators, plumbing and sewerage systems) contribute marginally to the operational energy costs of the building. Therefore, the first area to focus our energy conservation and efficiency efforts on is HVAC systems, followed by artificial lighting.

An argument has been expounded that as operational energy efficiency improves, these components of energy used in the skyscraper's life

Fig. 53   Delivered energy use in a typical air-conditioned office building (Source: Yeang)

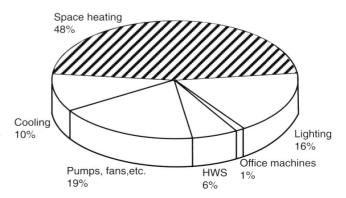

Space heating 48%

Cooling 10%

Pumps, fans, etc. 19%

HWS 6%

Office machines 1%

Lighting 16%

cycle will become less important than the embodied energy used in the construction of the skyscraper in the first place, which would become proportionally more significant (Lawson, 1996) (in chapter 4). However, unless the building approaches virtually zero energy consumption in its operational phase, it is likely that the 'initial energy costs' of the green skyscraper will remain small compared to the 'operational energy costs'. It has been said that the preferred standard for the annual primary energy consumption in Germany should be less than 100 kWh/m² per year (in Herzog, ed., 1996, p. 42).

The designer should approach operational systems in an order of priority based on the following (see Figs. 54 and 55):

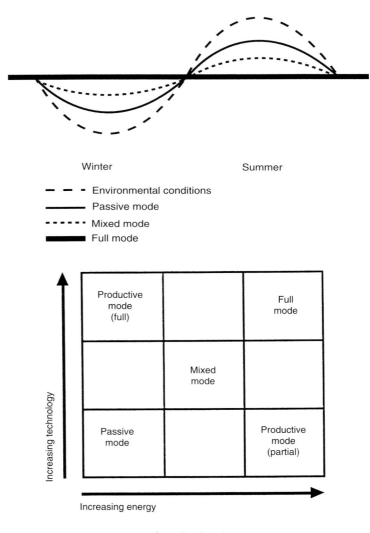

Fig. 54 Comfort ranges of different modes

Winter   Summer

- - - Environmental conditions
——— Passive mode
····· Mixed mode
▬▬▬ Full mode

Increasing technology

Productive mode (full)		Full mode
	Mixed mode	
Passive mode		Productive mode (partial)

Increasing energy

Operational modes

Fig. 55 Types of modes of operational systems

● Passive systems of energy use and recycling. These systems should be preferred over active systems (i.e. systems that use electro-mechanical devices), and such passive systems should be maximised.
● Mixed-mode systems (i.e. partially electro-mechanically assisted systems that optimise other ambient energies of the locality).
● Full-mode systems. These active systems should have low energy and low environmental impacts (e.g. low or even zero carbon dioxide emissions).
● Productive systems, being systems that generate on-site energy (e.g., photovoltaic systems).

Generally, the management of the building's operational systems (whether passive, active, mixed-mode or a composite) affects the overall thermal and ecological performance and impact of the building. We should note that the full-mode (active) systems have the biggest effect on the overall energy consumption in the skyscraper's or other large building's lifetime use of energy. But in order to be effective in achieving real energy conservation, all three aspects should be taken into account – beginning at the design and planning stages.

It is not possible within the scope of this treatise to cover the applications of all the different modes for all climatic zones and for all building types. Different building uses (e.g. whether residential or office or retail, etc.) would have different systems related to their use and time of use during the day.

Maximisation of Passive-Mode Systems

Our first design step to take is to look at the range of options for passive design based on the climatic conditions of the site locality and to maximise these opportunities. Priority has to be given to passive systems over active and mixed-mode systems (see Fig. 56) because this is the way to achieve the ideal level of servicing for ecological design (see chapter 4) because it represents the lowest level of consumption of energy from non-renewable sources. Passive design is essentially low-energy design achieved not by electro-mechanical means, but by the building's particular morphological organisation. Passive systems use various simple cooling and/or heating techniques to enable the indoor temperature of the building to be modified through natural and ambient energy sources in the natural environment. Obviously, the range of opportunities depends on the latitude of the location of the project site.

Passive comfort measures	Active comfort measures	Ice Caps	Tundra	Uplands	Continental	Temperate	Mediterranean	Subtropical	Tropical	Savannah	Steppes	Desert
Natural ventilation		○	○	①	④	⑥	⑥	⑦	⑦	⑦	⑦	⑦
	Mechanical ventilation	⑤	⑤	③	③	③	④	⑤	⑥	⑥	⑥	⑥
Night ventilation		○	①	②	③	⑤	⑥	⑦	⑦	⑦	⑦	⑦
	Artificial ventilation	○	○	○	①	①	③	⑤	⑤	⑤	⑤	⑥
Evaporative cooling		○	○	○	①	②	③	②	②	⑤	⑥	⑦
	Free cooling	○	○	○	④	③	⑤	⑥	⑥	⑦	⑦	⑦
Heavy construction		③	④	④	⑥	⑤	⑥	②	②	③	⑤	⑥
Lightweight construction		③	③	②	②	③	③	⑤	⑤	⑥	④	④
	Artificial heating	⑦	⑦	⑦	⑦	⑥	④	○	○	②	④	①
Solar heating		②	③	⑥	⑥	⑦	⑥	○	○	②	③	○
	Free heating	⑦	⑦	⑦	⑥	⑥	⑤	○	○	○	③	○
Incidental heat		⑥	⑥	⑥	⑤	⑤	④	○	○	①	②	○
Insulation/permeability		⑦	⑦	⑦	⑦	⑥	⑤	○	○	①	③	④
Solar control/shading		○	①	③	④	⑤	⑥	⑥	⑥	⑥	⑦	⑦
	Artificial lighting during daytime	⑥	⑥	④	④	④	③	③	③	②	②	②
Daylight		⑥	⑥	⑥	⑥	⑥	⑥	⑤	⑤	⑤	④	④

○ ① ② ③ ④ ⑤ ⑥ ⑦
No importance — Very important

Fig. 56 Energy saving measures by global regions (Source: Lloyd Jones, D., 1998)

Strictly considered, passive systems should exclude any electro-mechanical devices that use non-renewable forms of energy. Climate responsive design (i.e. bioclimatic design) thus reduces consumption of non-renewable energy, mimimising such mechanical systems. Although there are some who might argue that the use of a fan or pump may be permitted within this category (e.g. in Givoni, B., 1994, p. 1), it is contended here that full 'passive' systems mean operational systems without the use of any electrical-mechanical devices or systems. 'Active' means energy dependent full-mode systems. Mixed-mode means partially energy dependent or partially assisted electro-mechanical systems.

In temperate climates, bioclimatic principles seek to reduce heat gains by conduction, radiation and convection through walls and

windows in the summer, and to reduce heat losses towards the
exterior in winter. Passive cooling systems transfer incident energy
to natural energetic deposits, or heat sinks, such as the air, the
upper atmosphere, water and earth. Passive warming systems store
and distribute solar energy without the need of complex controllers
for distribution. Applying passive systems, it is possible to reduce
the average external temperature values in summer and increase
indoor temperatures in winter.

The following are some of the passive methods to be used in the
design of the green skyscraper and other intensive building types,
with the remaining energy needs to be met by active systems or
mixed-mode systems powered by ecologically sustainable forms of
energy:
● Built-form configuration and site layout planning.
● Built-form orientation (of main facades and openings, etc.).
● Facade design (including window size, location and details).
● Solar-control devices (e.g. shading for facades and windows).
● Passive daylight devices.
● Built-form envelope colour.
● Vertical landscaping (i.e. use of plants in relation to the built-
form).
● Wind and natural ventilation.
These methods are discussed below.

Passive Design by Building Configuration

We need to shape the form of the skyscraper or the large building
type in relation to the energies of the ambient environment and
the meteorological qualities of the locality to achieve passive
response (see Fig. 57). However, reducing the heating energy
requirements (e.g., by optimising the incoming heat) is not simply a
matter of building orientation; it is also influenced by the form of
the building and the ratio of volume to surface (see Daniels, 1995).

The skyscraper's (and other large intensive building's) form and
components need to be shaped and placed in such a way on the site
as to function in a low-energy way. This approach has been well
developed by others for the low-rise and medium-rise building
types, but less so for the high-rise building type (e.g., in Olgyay,
1993). If the building has not been configured or orientated to maximise its passive benefits, then any electro-mechanical active systems and devices that are later installed may have to 'correct' for
some of the earlier design 'errors' – which makes total nonsense of
the goal of energy efficiency design.

Fig. 57   Optimum aspect ratios of buildings (Source: Yeang, K., 'Bioclimatic Skyscrapers', Ellipsis, 1994)

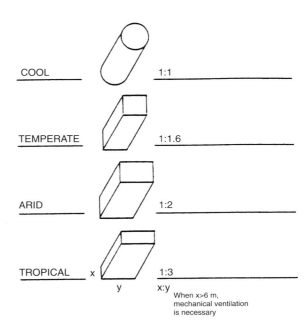

COOL _____ 1:1 _____

TEMPERATE _____ 1:1.6 _____

ARID _____ 1:2 _____

TROPICAL   x _____ 1:3 _____

y   x:y

When x>6 m, mechanical ventilation is necessary

FORM

Diagram shows the optimum aspect ratios of buildings in each climate zone, the best orientation of main facades, and the distribution of primary mass to achieve maximum solar shading or solar gain respectively. Research has shown that the preferred length of the sides of the building, where the sides are of length x:y, is:

- Cool zone   1:1
- Temperate zone   1:1.6
- Arid zone   1:2
- Tropical zone   1:3

Looking at the aspect ratios, we see that lower latitudes require an elongated from, to minimize the east and west exposure. This form gradually evolves into a 1:1 ratio, i.e. to a cylindrical form, as we reach the higher latitudes in the north, where the surface capable of utilizing solar gain should be as large as possible.

It is generally held that the built form should have a 1:2 to 1:3 length ratio, for climatic zones nearer to the equatorial zone, and lesser at the higher latitudes (see Fig. 58), so that it is twice as long as it is wide (Cole, 1995). This provides the added advantage of reducing shading impact on any buildings located to the south (if the building is located below the equator). We should ensure that the long axis of the built form is oriented east-west so that the long length of the building faces north-south. This enables the majority of the windows to be designed into the north wall for sites at the equator (and vice versa if above the equator) so that sun penetration into the building will be maximised.

The designer must see the internal and external interfaces of the building's form, fabric and comfort in relation to the site, climate and global ecological impact of the building (in Pearson, D., 1989). For instance, in the hot-humid tropical zones, one approach is to shape the skyscraper's or large building's floor plate, built form and service core in relation to the sun's path in the locality's latitude, in order to create a solar buffer to reduce solar penetration into the inside of the building. It is evident that tall buildings are more exposed than others to the full impacts of external temperatures, winds and sunlight. built form configuration, orientation, floor-plate shape and use of buffer components can have important effects on energy conservation and natural lighting to interior spaces.

The building's service core placement also determines which parts of the floor plate's periphery have openings (e.g., for ventilation and views); their location can benefit the building's thermal

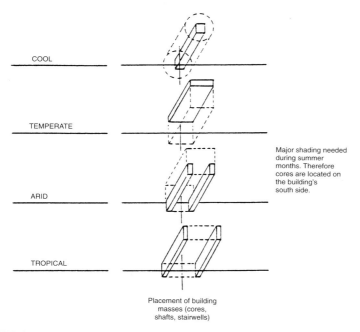

COOL

TEMPERATE

Major shading needed during summer months. Therefore cores are located on the building's south side.

ARID

TROPICAL

Placement of building masses (cores, shafts, stairwells)

VERTICAL CORES AND STRUCTURE

The arrangement of primary mass can be used as a factor in bioclimatic design as its position can help to shade or retain heat within the building form.

The cool zone requires the maximum perimeter of the building to be open to the sun for heat penetration. Therefore, the primary mass is placed in the centre of the building so as not to block out the sun's rays and to retain heat within the building. The arrangement of the primary mass in the temperate zone is on the north face, so as to leave the south face available for solar heat gain during the winter.

In the arid zone, the cores should also be located on the east and west sides, but with major shading only needed during the summer. Therefore, the cores are located on the east and west sides, but primarily on the south side.

For the tropical zone, the cores are located on the east and west sides of the building form, so as to help shade the building from the low angles of the sun during the major part of the day.

Fig. 58   Vertical cores and structure (Source: Yeang, K., *Bioclimatic Skyscrapers*, Ellipsis, 1994)

performance as well. The designer must consider the sun path and the winds of the locality in aggregate with other factors in making design tradeoffs (e.g., direction of best views, site shape, neighbouring buildings; see Fig. 58). The service cores can be positioned on the 'hot' east or west sides of the building, or both, to serve as solar buffers in the tropical zone. Studies show that significant savings in air-conditioning can be achieved from a double-core configuration with window-openings running north-south and cores on the east and west, even in temperate and cold climatic zones. This is most applicable at the equator and at latitudes up to about 40° above the equator. This placement prevents heat gain into the internal user spaces and provides a 'spatial thermal insulation' buffer to the hot sides, while at the same time maximising heat loss away from user spaces. In cold and temperate climates, in the upper or southern regions of the biosphere, they could serve as 'wind buffers'.

Of the various possible service-core positions (i.e. 'central core', 'double core' or 'single-sided core'), the double core is to be preferred (see Fig. 59). The benefits of a peripheral service core position are:

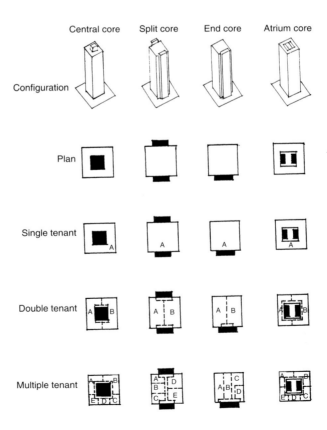

Fig. 59 Service core configurations (Source: Yeang)

	Annual cooling load kWhr/m²/annum				Average cooling load
	N (S)	NE (SW)	E (W)	SE (NW)	
Centre core	166.4	170.8	167.4	170.1	168.7 137%
Split core	121.4	124.5	123.6	123.3	123.1 100%
Side core (opposite site: wall)	123.3	125.4	122.4	124.5	
Side core (opposite site: wall)	124.5	128.4	127.4	127.9	125.5 102%

Fig. 60   Orientation, core position and cooling load (Source: Nihon Sekkei, 1984)

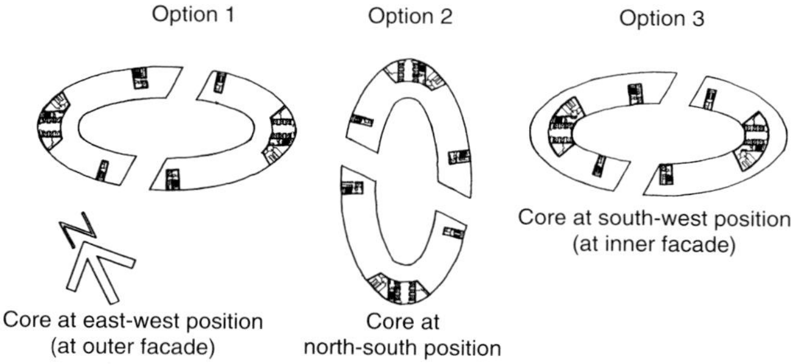

Notes:
Location   : Tokyo (Latitude 36 °N)
Typical floor area : 2,400 m²
Floor heigh   : 3.7 m
Window-wall ratio   : 60 %

Lighting   : 30 W/m²
Infiltration   : 1 change/h
People   : 7 m²/p

Temp./humid
Cooling   : 26 °C 50 RH
Heating   : 22 °C 50 RH
Air cond. floor
area ratio   : 65 %
Outdoor air intake : 4.5 m³/m²h
Length–breadth   : 1:1.5
Insulation   : Foam
                      polystyrene 25 mm

Option 1

Core at east-west position
(at outer facade)

North	= 33.8
East	= 43.1
South	= 34.9
West	= 43.4

Total OTTV (W/m²) = 38.8
(90%)

Option 2

Core at
north-south position

North	= 34.2
East	= 48.6
South	= 35.0
West	= 47.6

Total OTTV (W/m²) = 41.4
(96%)

Option 3

Core at south-west position
(at inner facade)

North	= 35.3
East	= 50.2
South	= 36.0
West	= 50.3

Total OTTV (W/m²) = 42.9
(100%)

Assumption: Adequate sunshading on all relevant facades

Fig. 61   Passive mode optimisation by building configuration and orientation. This diagram shows the OTTV (Overall Thermal Transmission Value) of alternative configurations of elevator cores and orientations of the built form. (Source: Yeang)

- No fire-protection pressurisation duct, resulting in lower initial and operating costs
- A view out with greater awareness of place for users
- Provision of natural ventilation to the lift lobbies (and thus further energy savings)
- Provision of natural sunlight to lift and stair lobbies
- A safer building in event of total power failure
- Solar buffer effects and/or wind buffer effects

On sites that are above the equatorial belt, the service-core placements might be adjusted in accordance to the solar path, which would contribute to shaping the floor plate. By adopting the above strategy, the structure's configuration would be shifted away from the conventional central-core configuration of most skyscrapers.

The consequences are summarised in figure 60, which shows the correlation of different core locations and orientations with the annual cooling load. The core type that provides the minimum air-conditioning load is clearly the double-core configuration, in which the opening is from north to south and the core runs from east to west. Conversely, the core type characterised by maximum air-conditioning load is the centre-core configuration, in which the main daylight opening lies in the southeast and northwest directions (see Fig. 61).

Generally, we can conclude that a skyscraper arranged longitudinally from north to south has to bear an air-conditioning load that is 1.5 times in the theoretical conditions (i.e. without special facade treatment for the external walls) that of a building arranged longitudinally from east to west (see Fig. 58).

Passive Design by built form Orientation

This strategy relates the shape of the building's floor plate, its position on the site and its orientation to the sun's path and the wind direction of the locality (see Fig. 62). For instance, in responding to the sun path in the tropical zone, the building's shape should be rectangular along the east-west axis to reduce solar insolation on the wider sides of the building. This is because the greatest source of heat gain is solar radiation through the windows. The gain, of course, varies markedly with the time of day and the angle of incidence.

We must be careful to distinguish between true north (or solar north) and magnetic north, which are not the same. True north varies through time but is approximately 11° west of magnetic north. When designing, we must ensure that orientation is related to true

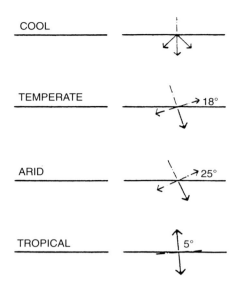

COOL

TEMPERATE → 18°

ARID → 25°

TROPICAL 5°

Fig. 62   Climate influences on built form and orientation and mass distribution (Source: Yeang, K., *Bioclimatic Skyscrapers*, Ellipsis, 1994)

ORIENTATION

Orientation as well as directional emphasis changes with latitude in response to solar angles.

Zone	Building's main orientation	Directional emphasis
Cool	on an axis facing south	facing south
Temperate	on an axis 18° north of east	south-south-east
Arid	on an axis 25° north of east	south-east
Tropical	on an axis 5° north of east	north-south

The optimum orientation of a building and the placement of the main facades become clear when we study this diagram. Orientation is an important factor in 'bioclimatic planning' since directional emphasis can help keep heat in or out of a building.

north because misunderstandings can seriously affect shadow diagrams.

If the site does not align with the sun's geometry on its east-west path, other building elements such as the building's service cores can follow the geometry of the site to optimise the effects on column grids, basement car-parking layouts and other features. However, other features may need to be introduced to correct for the hot facades.

Some might assume that the square building floor plate will have a lesser peripheral area and will be less exposed to the effects of external air than other built form shapes, resulting in lower air-conditioning load. Studies show, however, that it is actually the circular floor plate that has the least surface exposure. But despite this shape's better surface exposure, it is the rectangular shape that is actually better for solar control (see Fig. 63).

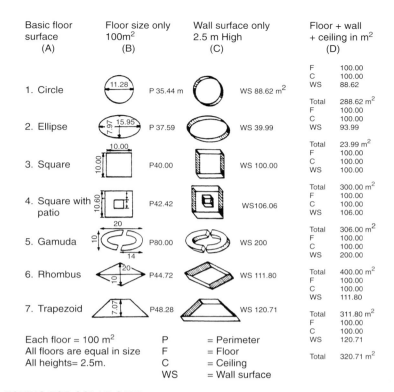

Fig. 63 Surface exposure for different built forms (Source: Germany, 1983, adapted)

	Basic floor surface (A)	Floor size only 100m² (B)	Wall surface only 2.5 m High (C)	Floor + wall + ceiling in m² (D)
1. Circle	11.28	P 35.44 m	WS 88.62 m²	F 100.00 C 100.00 WS 88.62 Total 288.62 m²
2. Ellipse	15.95 / 7.97	P 37.59	WS 39.99	F 100.00 C 100.00 WS 93.99 Total 23.99 m²
3. Square	10.00 / 10.00	P40.00	WS 100.00	F 100.00 C 100.00 WS 100.00 Total 300.00 m²
4. Square with patio	10.60	P42.42	WS106.06	F 100.00 C 100.00 WS 106.00 Total 306.00 m²
5. Gamuda	20 / 10 / 14	P80.00	WS 200	F 100.00 C 100.00 WS 200.00 Total 400.00 m²
6. Rhombus	20 / 10	P44.72	WS 111.80	F 100.00 C 100.00 WS 111.80 Total 311.80 m²
7. Trapezoid	7.07	P48.28	WS 120.71	F 100.00 C 100.00 WS 120.71 Total 320.71 m²

Each floor = 100 m²
All floors are equal in size
All heights= 2.5m.

P = Perimeter
F = Floor
C = Ceiling
WS = Wall surface

ZONING FOR SOLAR GAIN
Diagram indicates the location of spaces that can be used for solar heat gain. The location follows the sunpaths in each climate zone; in the tropical and arid zones, these are on the east- and west-facing sides; in the temperate and cold zones, they are on the south-facing side.

Provided passive solar energy use is appropriate, and will not result in overheating, we need to site the building to allow for use of solar gains to offset heating requirements in temperate and cold climatic zones. This is likely to be appropriate in buildings with low occupant density, because high internal heat gains from equipment will be less likely.

Above the equatorial zone, in temperate and cold climatic zones, there are two basic strategies in terms of shaping the built form to achieve minimum energy impact:

● Minimise the surface area-to-volume ratio, design to high insulation levels and compact the building form to minimise heat losses in winter conditions. This strategy helps minimise both the building materials consumed and the direct energy requirements in fuel terms.

● Reduce plan depth to maximise the opportunity to use natural ventilation and natural daylight.

Current thinking in low-energy office-use skyscraper design favours the second of these two options as the preferred passive solution whenever possible. It allows maximum use of natural ventilation and daylight to minimise energy consumption. This approach can be applied even if at first glance the scope for it appears to be limited. For example, when planning internal spaces, the designer must ensure that rooms without specific environmental requirements can benefit from natural light and ventilation. General offices can be located to make use of the external ambient energy resources available, while conversely, areas with high heat gains (such as kitchenettes and computer rooms) should be sited to the north (if the site is south of the equator), or in the central core of a deep-plan building if they will require to be fully mechanically treated (e.g. clean rooms and some computer facilities). In other words, the design strategy is to make use of external ambient energy in spaces where the gain will be of benefit and avoid solar gains and natural ventilation only if this will exacerbate an existing problem, or if it is completely inappropriate to the task (see Fig. 64).

The first decision taken in relation to the orientation of the building on the site can affect every other later decision. Every

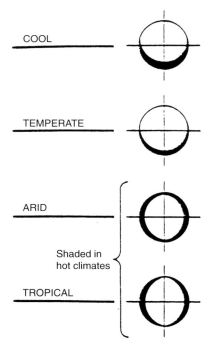

COOL

TEMPERATE

ARID

Shaded in hot climates

TROPICAL

Placement for solar gain, especially in arid and tropical zones; these areas must be shaded.

Fig. 64   Influences on built form (Source: Yeang, K., *Bioclimatic Skyscrapers*, Ellipsis, 1994)

building site is unique and therefore the design of the intensive building type is exclusively related to that site. This immediately places constraints on the ensuing design decisions. When deciding how to make an environmentally responsive intervention with a new building within the site, two major site factors must be addressed: the local climate and the environmental impact of the building on the site and vice versa. Local climate considerations can be positive and negative. The building's orientation can take advantage of free energy from the sun in terms of both heat and light. Wind effects can be mitigated by shelterbelt planting or permeable walling, or can be used (in combination with compatible devices) for natural ventilation. By building up a picture of basic microclimatic information, it is possible to identify the most suitable siting and configuration for the building, eliminating unsuitable (polluted, overshadowed) areas of the site and using the potential of the remaining land to the fullest through the building form, planting and shelter belts. Investigative studies include using overlay techniques for clarification.

The final design should incorporate adequate fenestration to make use of free solar energy and natural daylight while optimising heat losses and avoiding glare. This will seldom be achieved by glazing ratios (the ratio of glass to solid surfaces in the facade) of ideally above 20 percent if single-glazing (particularly in the temperate zone), but can be checked by manual calculation or computer modeling.

Recent advances in external wall systems include the use of special glass, double-glazing, composite double-glazing with blinds, double facades, etc., which enable higher glass-to-solid ratios.

Our intensive building should also be considered in relation to other skyscrapers and buildings on the site and within the block. Aspects such as aesthetics, overshadowing, self-shading, climate variations, vegetation and pollution should be examined to avoid negative effects on existing and new buildings. An overall site strategy for energy use and an integrated energy policy should be developed at an early stage; this will include consideration of the use of waste heat, potential to generate electricity on site using a combined heat and power scheme (i.e., cogeneration) and renewable energy sources (e.g. wind turbines, photovoltaics, etc.). In other words, the site should be considered holistically and not in isolation. In shaping the floor plate and the facade design, the designer might also identify key 'view corridors' from the building to its surroundings that need to be maintained.

In temperate climates where the sun's path is lower, adequate measures have to be taken to avoid excessive overheating in the summer while ensuring that the potential of useful winter solar gains is maximised. In temperate climates, the sun has to be designed 'out' as well as 'in'. A balance between heat losses from the north and heat gains to the south must be achieved. This can be assisted by careful material selection as well as space planning. For example, heat gains from the M&E plant rooms can reduce heating requirements in peripheral zones. Another strategy is for the service cores and circulation spaces, stair wells and corridors to be used as buffers between the accommodation areas and the outside, either by siting them on the north to help reduce heat losses, or by reducing excess gain to the south (if designed, for example, as walkways and galleries).

In temperate zones, solar gain will be of benefit in most buildings early in the morning when the building is cold. This is particularly true in non-domestic buildings, where use of solar energy in east and southeast zones in winter can provide free early morning pre-heating to offset heating loads. The associated risk of summer overheating can be addressed by solar shading (see below).

In situations where internal heat gains are high, solar gains to rooms on southwest- and west-facing aspects may contribute to overheating, for during much of the year sunlight will be incident on these facades later in the day when the interior is already warm. Contrary to current recommendations for residential use skyscrapers such as apartments and hotels, where north-facing glass should be minimised, in non-residential skyscrapers north-facing glass should be optimised to make use of daylight while avoiding excessive heat losses. It is important to be aware that artificial lighting can account for up to 50 percent of the overall electricity costs in a modern office skyscraper. Coupled with excess uncontrolled solar heat gain, this can result in unnecessary energy expenditure on air conditioning – which could have been avoided by a clearer understanding of the cumulative impact of internal and external factors.

Passive Design by Facade Design

Our next task, facade design, should be given greater priority over the building's contents (i.e. its operational systems) and should in effect be designed in combination with the optimisation of the passive systems, mixed-mode and such active systems as are used. The permeability of the skin of the building to light, heat and air and its visual transparency must be controllable and capable of modification, so that the building can react to changing local climatic conditions. These variables include solar screening, glare protection, temporary thermal protection and adjustable natural ventilation options. A well-designed building envelope will yield significant energy savings (see Fig. 65). The building is like our 'third skin', after our physical skin covering and our clothes; all these layers of enclosure need to function naturally and in harmony with our bodies and the natural environment. By analogy, the building's facade as our 'third skin' needs to breath and function as a regulator, protector, insulator and integrator with the natural environment.

Thus, the ideal external wall should be an environmentally responsive filter. The envelope should have adjustable openings that operate as sieve-like filters with variable parts to provide natural ventilation, control cross-ventilation, provide a view outward, give solar protection, regulate wind-swept rain and discharge heavy rain, provide insulation during cold seasons (and in temperate zones, meet the demands of a hot summer, a cold winter, and two mid-seasons), and enable a more direct relationship with the external environment. In temperate zones and above, the designer

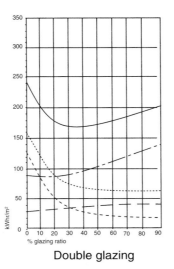

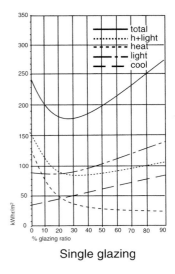

Figs. 65 Annual primary energy consumption (lighting and heating for south-facing office) (c. 50° Latitude) (Source: Baker, 1992)

Double glazing

Single glazing

should consider the angles for both summer sun penetration and winter sun penetration, as these differ (see Fig. 65). The green approach runs contrary to those facade designs that use hermetically sealed skins. The 'green' facade has to be multi-functional in its design. It can reduce solar heat gain to the space through external shading devices, provide fresh air ventilation, serve as an acoustic barrier, give maintenance access and make a contribution to the building's aesthetics.

The double-layered facade system operates on the principle of using a ventilated double 'skin' with an intermediate shading device. The intermediate shading device reflects out a majority of the incoming solar radiation back through the external glass. The proportion of absorbed solar radiation is converted into 'sensible' heat and radiated back into the air space between the inner and outer glass units. Ventilation of the heat in the air space is dependent on the effects of external wind pressures and/or the 'stack' effect. The 'stack' effect works on the principle that the heat absorbed and re-radiated by the blinds and glazing rises within the cavity. Cooler air is drawn into the air space to replace the buoyant warmer air, which is ventilated; thus, an air stream is produced to ventilate the solar heat gains. The system is most efficient when external wind pressure supplements the ventilation effect, but since the occurrence and intensity of wind is variable, optimum performance of the design also varies (see Fig. 66).

For the stack effect to work effectively, the depth of the air space between the layers must be greater than 250–300 millimetres and

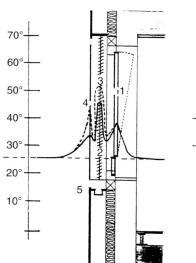

1. Combined tip-tilt sash
   Wood/aluminium
   Double glazing

2. Ventilated cavity

3. Louvre

4. Single glazing as
   weather protection

5. Air inlet

——— Resulting temperatures
     with ventilation due
     to wind pressure
- - - - Resulting temperatures
     with calm condition
     (natural stack effect only)

Outside air temperature
T = 25 °C

Incident solar radiation
qs = 800 W/m²

Air change rate
V/A =50 m³/hm²

Fig. 66   Naturally ventilated wall (double wall) (Source: Permasteelisa, 1998)

the external vent dimension must be at least 150mm. This results in a considerably deep and heavy facade structure. While the system is effective in controlling solar heat gains, the introduction of cold external air into the cavity during winter means that the benefit of an air 'buffer' is negated. Natural ventilation due to external wind pressures must also be carefully considered, as tests have shown that strong oscillations of the blinds were noted at external wind speeds of around 30 metres per second. This is an important consideration in countries subject to high maximum wind gusts. However, the benefits of a double facade include dampening of noise emissions, reduction in high-wind pressures in skyscrapers and natural ventilation. However, wind pressure reduction within a facade also reduces the air exchange on calm days. The possibility of creating a greenhouse effect in front of each window can also moderate the outside air in winter, and creates a heat trap in summer. This enables the use of an outside sun shading that is not subject to wind forces, so that tall buildings with clear glazing can be built without risking excessive solar gains.

Active and interactive facades operate on the same principle as a naturally ventilated double-skin wall, but have a number of advantages. As with a naturally ventilated wall, the 'sensible' heat built up within the cavity is extracted by forced or controlled ventilation. The wall is more compact, requiring a much smaller overall section depth than a naturally ventilated wall. With an active wall, the heat gains are extracted using internal room air. In temperate and cold climatic zones during winter, the inner glass may be kept closer to room temperature (within 1–2° C), thus eliminating any cold radiation effects and allowing greater use of the perimeter area of the office building. The heat energy removed by an active facade also can be used in conjunction with a heat exchange system to provide further energy savings. An interactive wall is similar to a naturally ventilated facade, but the rate of ventilation is controlled via the use of a small energy-efficient fan powered by solar energy or by conventional electrical means. This system is a 'stand alone' system independent of the main building mechanical systems.

Insulation

Another effective way to reduce energy consumption in the building, particularly in temperate and cold climates, is to increase external-wall insulation to reduce leakages and to lower the ratio of solid to glass area. Insulation reduces the rate that heat can flow through elements in which it is installed. In heated and cooled buildings, this will result in significant energy savings and thermal comfort.

External wall surfaces having direct solar insolation should be insulated and the 'time lag' taken into consideration. External materials used might be those that are effective heat sinks (e.g., aluminium cladding), or be designed with a double-layered (or even triple-layered) ventilating space. The thermal transfer properties of the window frame and glazing systems should be verified. For instance, aluminium framing fitted with thermal bridging can assist in reducing heat transfer. Current standards of external-wall insulation in temperate climates are around 0.45 Wm^2 K. Design efforts should therefore try to achieve insulation levels lower than this. By reducing or using several times the insulation level, the carbon dioxide emissions could also be reduced (e.g., Vale, B., and Vale, R., 1991); a figure for carbon dioxide emissions of only 28 Kg/m^2 per year has been achieved in residential design by using three times the recommended levels of insulation. A single glazed window has a heat loss rate over 10 times greater than the United Kingdom Building Regulations (5.7 Wm^2 K for a wall element). Double-glazing (depending on the air gap may reduce this to 2.8 Wm^2 K, and triple glazing may reduce this to 2.0 Wm^2 K. This might further be lowered by reducing the solid to glazing ratio of external walls.

It is often presumed that higher insulation levels in the facade and roof (although minimal in the case of the skyscraper) are the obvious solution to reducing energy use. However, more insulation is not necessarily the answer. The thermal performance of materials has to be considered in conjunction with other factors, among which are the following:

Air leakage. In the temperate and cold climatic zones, the loss of heat through air leakage can begin to dominate 'heat loss' if not addressed simultaneously with insulation levels.

Heating system. Similarly, the heating system and components must be designed and selected with effective energy use in mind and must match the building fabric and purpose. A well-insulated building with an inefficient or poorly controlled heating system will probably overheat – and without saving energy. Potable hot water production may also become a critical factor in the overall energy picture and so a holistic approach is needed.

Summer and winter. In the temperate and cold climatic zones, high insulation levels will reduce the rate of heat loss in winter and summer. They also reduce the potential for heat gain. The thermal performance must be evaluated to optimise heat gains and losses to avoid trapping heat from internal sources in summer while avoiding excessive losses in winter. Simultaneously, potentially useful winter gains should not be eliminated by only considering the avoidance of summer gains. The optimum solution will be achieved by optimising insulation levels, designing heating systems for high operating efficiency and providing solar protection in summer without compromising winter solar gains.

The above strategy does not, of course, exclude the benefits of high insulation standards but points out the fact that solutions should be carefully evaluated. It is possible to achieve insulation values of well above the statutory requirements at little additional cost, using modern techniques and materials, and this approach should be adopted.

Passive Design through Solar-Control Devices

Generally, on the 'hot' sides of the building (the east and west sides), regardless of the latitude, some form of solar shading is required, making due allowance for glare and the quality of light entering the spaces (see Fig. 67). Full-height glazed curtain wall may be used on the non-solar facades, provided it meets the other criteria (e.g., insulation values; see Fig. 68). The westward solar-facing wall has the highest intensity at the hottest time of the day. The correct form of angle control (through computer-aided design) is held to be the key to the optimisation of a facade, since direct radiation of internal areas should be avoided for lengthy periods during the year, while maintaining high daylight utilisation (radiation transmission) (in Daniels, 1995). Depending upon the season and time of day, the angle control of the louvres achieves optimal daylight incidence in combination with minimal heat gain (see Fig. 69). 'Intelligent facades' operate with automated angle control, regulated by incident radiation and outside air temperature. During periods of unwanted outside thermal gain, the louvres are positioned at a steeper angle than during periods when passive solar gain can be taken advantage of to reduce the energy costs for heating rooms. Angle control is thus an important factor with regard to total solar energy transmission. Another is the type of glazing used between the indoor space and the air cavity separating it from the outside.
In temperate climates, designers should reduce the need for air conditioning in the summer and the need for heating in winter. The strategy is to 'extend' the middle seasons (spring and autumn) through enhanced cross-ventilation in the summer, late spring and

Energy reduced by shifting up indoor temperature and providing sun's protection on east and west glass walls

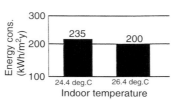

Energy reduced by shifting up the indoor temperature

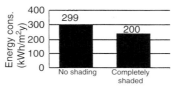

Energy reduced by the provision of shading devices

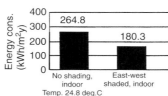

Fig. 67   Impact on energy by lowering internal temperature and by shading (Source: Karyono, 1996)

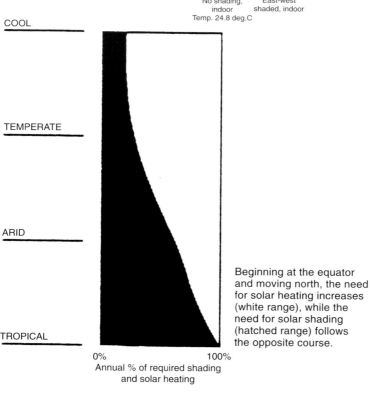

COOL

TEMPERATE

ARID

TROPICAL

0%   100%
Annual % of required shading
and solar heating

Beginning at the equator and moving north, the need for solar heating increases (white range), while the need for solar shading (hatched range) follows the opposite course.

■ shading
□ solar heating

Fig. 68   Annual percent of required solar shading and solar heating (Source: Yeang, K., *Bioclimatic Skyscrapers*, Ellipsis, 1994)

Fig. 69   Shading of windows by internal and external louvres (Source: Baker, 1992)

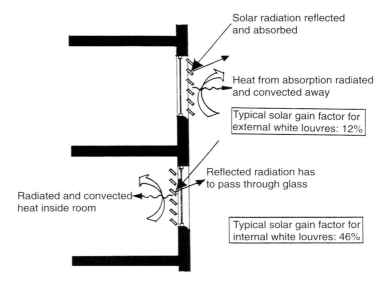

Solar radiation reflected and absorbed

Heat from absorption radiated and convected away

Typical solar gain factor for external white louvres: 12%

Reflected radiation has to pass through glass

Radiated and convected heat inside room

Typical solar gain factor for internal white louvres: 46%

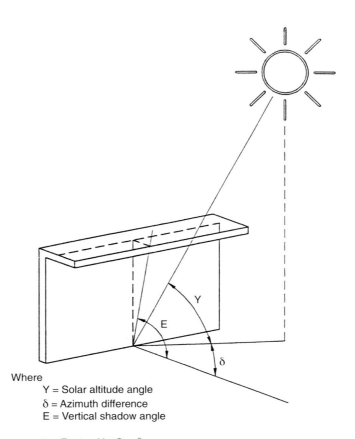

Where
Y = Solar altitude angle
δ = Azimuth difference
E = Vertical shadow angle

tan E = tan Y × Sec δ

Fig. 70   Solar sun-shade geometrical design (Source: Yeang)

early autumn. In the winter, we need to capture as much solar benefits as possible and reduce the need for heating the internal spaces through enhanced external-wall insulation (e.g., movable shields).

The most efficient form of solar control is to provide external shading devices over clear glass. Clear glass is often preferred, as it gives a more direct visual and natural relationship between the inside and outside of the building. Recent advances in glass technologies have produced "door" glass that give good light transmission but have lower shading coefficients.

The particular design can only come from understanding the solar geometry for the locality. Consider the sun-path diagram below for 0° latitude. In the solar geometry diagram in figure 70, E is found from the formula: $\tan E = \tan y \times \text{Sec } \delta$. Considering March 21 at 9:00 and 11:00 respectively yields the following calculation: solar azimuth angle a = 90°, wall azimuth angle w = 56°, hence $\delta = 90° − 56° = 34°$. The vertical shadow angles are calculated to be E= 49° and = 76°, respectively. Constructing these angles on the facade cross-section determines the effectiveness of the solar-shading devices.

Generally stated, solar heat-gain through windows in the building can be reduced by sunshades, balconies or deep recesses (such as totally recessed windows) or skycourts (see Fig. 71). This shading

Insolation
The four diagrams provide an overview
of the sun's influence on local climates.

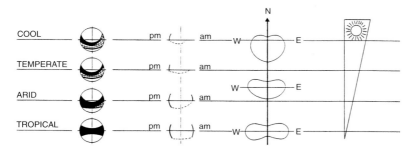

SOLAR PATHS REQUIRING SHADE	SUNSHADE ANALYSIS (VERTICAL AND HORIZONTAL)	INSOLATION	SUN REQUIREMENT DURING WINTER
The shading requirements depend upon the sunpath in each season. In the lower latitudes,there is a danger of overheating related to undesirable solar gain (darkly hatched range), whereas in the higher latitudes, overheating only occurs during the summer month. The hatched sections indicate the sunpath as observed in each climate zone.	Diagram indicates the optimum location of vertical sun shading (solid line), shielding buildings from low sun angles in the morning and evening, and of horizontal sun shading (broken line), blocking the high midday sun. Tropical zones require both types of shading throughout the year. In higher latitudes, horizontal and vertical shading is only needed during mid-seasons along east-, west-, and south-facing sides of buildings.	Diagram represents the shape of the sunpath in each climate zone. The sun or solar path becomes more southerly as one moves north, changing from a 'bow tie' pattern near the equator to a 'heart' pattern in the temperate zones.	Near the equator, seasonal variations are minimal and the need for solar heating is low, whereas, in higher latitudes, this need increases during the winter.

Fig. 71 Insolation (Source: Yeang, K., *Bioclimatic Skyscrapers*, Ellipsis, 1994)

cuts the huge solar heat gains that occur directly through window transmission and indirectly from hot surface conduction and radiation. Shading should also be designed to reduce glare as well as to enable the passive transmission of light into the deeper reaches of the floor-plate. This enables the designer to use clear glass and to give better daylight entry to internal spaces, which then reduces the lighting energy load as well as enabling a better aesthetic relationship between the inside and the outside of the building.

In temperate and cold zones, we need an adaptable facade to let the sun in through the windows in the winter months to heat the building in residential uses, and to reduce glare in office uses and to provide sun-shading in the summer to reduce heat gain. The building mass can be used to store thermal heat, but this is not possible in zones where the night air temperatures do not fall below comfort temperature to discharge the heat.

It is useful to provide external or mid-pane solar shading in preference to internal shading in situations where heat gain and glare are to be controlled. As movable devices can provide additional protection against heat loss in winter, it is suggested that application of an external device such as movable blinds or mid-pane protection is preferable to providing fixed shading (particularly on the east and west facades) with internal blinds. Combining energy, amenity and biological considerations, a combination of fixed and movable, solid and planted shading devices can be cost effective.

In the past, fixed shading devices have been preferred to movable ones, in part because of simplicity, low cost and minimum maintenance, but also because limiting human interaction in some ways limits room for error or misuse. However, they are not as effective as movable shades for anything other than to shade buildings facing due south. If it is the intention to employ a fixed solution, this should be borne in mind.

For example, in Britain, a horizontal shelf-type screen of approximately 0.7m width will be required to shade each metre height of exposed south-facing glass effectively in summer (from mid-May to early August) at a latitude of around 56°. Various configurations can be provided at the top of the glass, or a reduced depth of shelf can be installed lower down the pane. Alternatively, the shading device can be located partly inside and partly outside, providing a light shelf to throw light deeper into the space. Careful detailing is important to avoid excess glare or heat gain via an exposed upper pane. Louvred devices allow hot air to pass through and are not subject to snow or wind loadings. Architecturally, louvres can be used to provide articulation, and less projection is necessary to achieve an equivalent degree of protection compared with a solid shelf arrangement.

The east facade of the building requires solar shading as it is subject to the morning sun (see Fig. 72). The vertical shadow angle determines the effectiveness of horizontal solar-shading devices. The horizontal component of the sun's angle of incidence 'δ' is the difference between the solar azimuth angle and the wall azimuth angle. The solar altitude angle 'y' is read from the concentric circles, all of which are obtained from the sun-path diagram for the latitude of the building's locality for a given time of day.

Fig. 72 Types of shading devices having different effect on view and ventilation (Source: Baker, 1992)

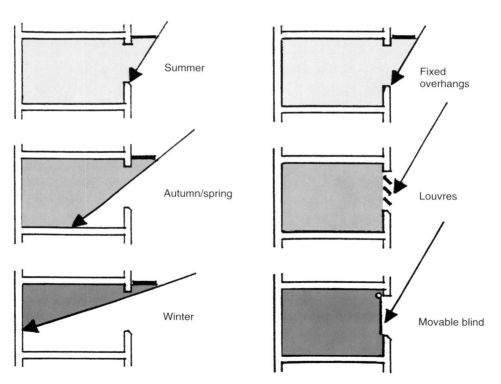

Fixed overhangs respond to solar elevation but do not synchronise well to actual heat demand and the need for daylight.

Fixed horizontal shading devices such as overhangs are very effective for south facade windows in higher latitudes but have a disadvantage in that they operate on a 'worst case' basis, depending on sun altitude only and not external temperature. Thus, these shades may block out useful solar gain under certain conditions when heating is required. East and west facing facades will always require some degree of vertical shading (see Fig. 73).

Movable shading has two main advantages: It can be adjusted to suit outside conditions to allow maximum benefit from the sun and to provide protection from glare and excessive heat gain; and, in winter, in temperate zones, devices can be closed to reduce heat loss from the building through radiation to the night sky. In terms

The sunpath diagram for the location of Chuang's Plaza, Kuala Lumpur, is used as a design tool when initially determining the form of the building and the areas which require least or most exposure to the sun. The building can now be described in terms of altitude and azimuth and by being plotted on the sunpath diagram, one can tell when the building will obstruct the sun or provide shade.

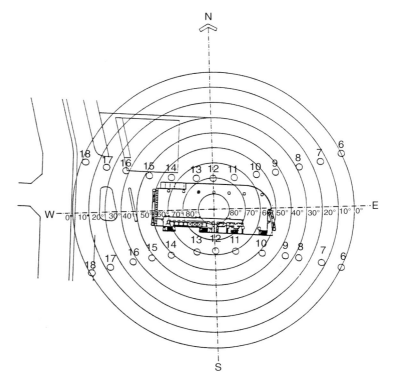

The sunpath diagram for the locality is used as a design tool when initially determining the building's form and those areas which require the least or the most exposure to the sun. The building can then be described in terms of altitude and azimuth. Using a sunpath diagram, one can tell when the building will obstruct the sun or provide shade.

Fig. 73   Sunpath diagram (Source: Yeang)

of protection from solar gain, external shading devices are more effective than mid-pane and more effective than internal devices. External devices prevent sunlight from entering the space, while internal devices allow solar energy to penetrate and then attempt to reflect the sunlight back out through the glass. The process is never 100 percent efficient, but the most effective solution will be achieved by a light-coloured or reflective finish on blinds or curtains. The external shading elements for the high-rise need also to be integrated with the facade cleaning strategy.

Tinted glass cannot be a substitute for sun shading. Tinting's best effect is to reduce thermal transmission to about 20 percent, which is still ineffective in attenuating conditions in hot climates and summer conditions in temperate climates with as much as 500–1000 W/m² levels of solar radiation. Tinting absorbs heat and so the external-wall gets hot, but the radiation heats the internal spaces and the air has to be made cooler for equal comfort. Hot glass further conducts the heat, so the cooling load is still higher. Worse, tinting cuts out daylight, affecting the lighting quality of internal spaces. Heat-absorbing glass absorbs shortwave (light) radiation and thus reduces heat gain to the inside of a building. However, depending on coincident internal and external climatic conditions, this energy stored as heat may be 're-radiated' into the space as the temperature outside begins to fall in the late afternoon. At this time, the internal equipment and casual gains tend to be at a peak. Thus, summer overheating can be more severe than would have occurred with clear glass, and winter heat gain may occur too late in the day to achieve heat absorption.

Studies have shown that the tints used can have two effects on building occupants:
● longwave (light) transmission is reduced, resulting in an increased need for artificial light, and/or larger window areas to achieve the same level of daylight;
● the psychological effect of looking at the outside world through brown, grey or green glass can be disturbing to occupants and has been suggested as a contributory factor in building-related health problems.

Solar-reflective glass can be used to reduce solar penetration without affecting the view to the same extent as heat absorbing glass. However, this solution reduces both shortwave (heat) and longwave (light) transmission, which results in reduced useful winter heat gain and year-round use of artificial lighting at times when natural light could have been sufficient – in other words, heat gain is eliminated at the expense of good quality natural lighting. Reflective glass can, however, be useful in situations where heat gain is not desired and where the use of external shading is not physically possible (particularly on west facades).

Low emissivity glass reduces direct heat gain by transmitting a greater proportion of light than heat. It reduces heat loss by re-reflecting heat back into the space and has an appearance similar to that of clear glass. It is thus useful for situations where daylight is required but solar heat gain should be minimised. It also allows the use of slightly larger windows for admitting daylight, without necessarily incurring an energy penalty in winter.

New 'intelligent' glazing systems, which overcome the problems of differing summer and winter requirements, are currently being researched and some are available already – at a price. The use of photo-chromatics, phase-change materials, holographic and electrically responsive glass and other technologies may become more commonplace. In the meantime, the environmentally responsive approach tends to encourage the use of clear or low emissivity glass, in high quality double or triple glazed units, wherever possible and with solar shading provided. Shading should preferably be by easily adjusted, external devices to allow occupant control. Failing that, fixed shading is effective on south facades and mid-pane shading is preferred to internal shading devices.

In the ecological approach the built system's external wall system has to be evaluated not only in terms of its ecological costs in terms of its inputs (e.g. embodied energy evaluation) but also designing for future recovery (see chapter 6).

Passive Daylight Concepts

Passive mode includes the use of passive daylight devices. The objective in the ecological design is to optimise the use of daylighting and to decrease the need for energy-consuming artificial lighting.

Most passive daylight techniques have worked to control incoming direct sunlight and minimise its potentially negative effect on visual comfort, i.e. glare, and the building's cooling load reduction by the reduction of heat gain. Direct sunlight, however, is an excellent interior illuminator when it is intercepted at the plane of the aperture and efficiently distributed throughout the building without glare. It is contended that windows and indoor spaces should be laid out in such a way that under overcast conditions (in temperate zones) with an external intensity of illumination of approximately 10,000 lux, a level of 200 lux can be achieved in the depth of rooms (see Herzog, ed., 1996, p. 43).

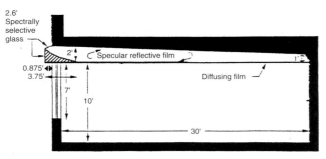

2.6'
Spectrally
selective
glass

2'
0.875'
3.75'
7'
10'
30'

Specular reflective film

Diffusing film

Section of trapezoidal light pipe design

Fig. 74   Light pipe
designs (Source: Lawrence
Berkeley National Labora-
tory, 1996)

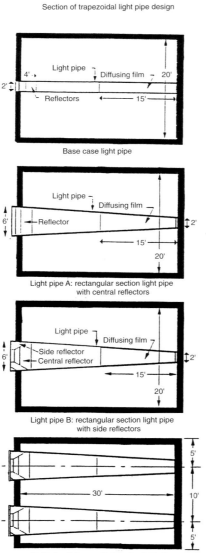

Light pipe
Diffusing film
4'
2'
Reflectors
20'
15'

Base case light pipe

Light pipe
Diffusing film
6'
Reflector
2'
15'
20'

Light pipe A: rectangular section light pipe
with central reflectors

Light pipe
Diffusing film
6'
Side reflector
Central reflector
2'
15'
20'

Light pipe B: rectangular section light pipe
with side reflectors

5'
30'
10'
5'

Light pipe C: trapezoidal section light pipe
with side reflectors (location
of two light pipes in space)

Transitional daylight designs can provide adequate daylight within about 4.6 metres (15 feet) of the conventional-height window. The use of larger windows and higher transmittance glazings to provide sufficient levels of daylight at distances further from the window has proven to be ineffective. Daylight levels increase asymptotically with distance from the window, so that a disproportionate amount of daylight/solar radiation must be introduced into the front of the room to achieve small gains in daylight levels at the back of the room.

There are a number of experimental perimeter daylighting systems that passively redirect beam sunlight further from the window using special optical films, an optimised geometry and a small glazing aperture and special glass (e.g. HOE, or holographic optical elements). The objectives of these systems are: (1) to increase daylight illuminance levels at 4.6 m to 9.1 m (15–30 feet) from the window aperture with minimum solar heat gains, and (2) to improve the uniformity of the daylight luminance gradient across the room under variable solar conditions throughout the year. Some other advanced systems use 'articulated light shelves' and 'light pipes' (see Fig. 74). Studies have shown that passive light-shelf and light-pipe designs can introduce adequate ambient daylight for office tasks in a 4.6 m to 9.1 m (15 to 30 feet) zone of a deep perimeter space under most sunny conditions with a relatively small inlet area; the light-pipe has been shown to perform more efficiently throughout the year than the light shelf.

The design of light-collecting systems relies upon the highly reflective and transmissive properties of the surface materials as well as their geometry to redirect sunlight more efficiently.

The advanced optical daylighting systems are based on the following concepts:

● By reflecting sunlight to the ceiling plane, daylight can be delivered to the workplace at depths greater than those achieved with conventional windows or skylights, and without significant increases in daylight levels near the window. This redirection improves visual comfort by increasing the uniformity of wall and ceiling illumination levels across the depth of the room.

● By using a relatively small inlet glazing area and transporting the daylight efficiently, lighting energy savings can be attained without severe cooling-load penalties from solar radiation.

● By carefully designing the system to block direct sun, direct source glare and thermal discomfort can be diminished. The challenge of the design stems from the large variation in solar position and daylight availability throughout the day and year.

A daylighted building, no matter how well designed, saves energy only if the daylighting can effectively and reliably displace electric lighting usage. Most daylighting designers agree that, in non-residential buildings, no amount of provision for convenient manual switching (even for 50 percent reduction of fixtures in use) will result in useful savings (Moore, F., 1996).

Energy savings in artificial lighting can begin by using narrow-width floor plates to reduce artificial lighting and to optimise natural lighting (e.g., at around 14m external wall-to-wall plate size). Earlier skyscrapers and large buildings with central cores have about 8.2 m (27 feet) depths from the external wall to the lift-core wall.

The external shading devices used on the facade can have an effect on the amount of natural daylight the building receives. Other devices include the use of overhangs from the floor above and external blinds. Provided the sunpath of the locality justifies their use, light-shelf devices can be used to reflect light back into the inner reaches of the floor space. These devices do not increase the quality of light but ensures a better distribution to the insides of the space. Detailed analysis would be required to ensure their effectiveness.

There is of course, an energy balance to be made between the savings in artificial lighting and the small increase in solar heat gain. We can also improve seating and work surface layouts, reduce glare, and have natural light through better windows and facade design. For more effective means, particularly in climatic zones with low sun-path, the use of holographic glass in the outer facade and the use of light shelves as scoops may extend natural light to 10 metres or more from the facade line (see Fig. 75). Studies have shown that access to sunlight and views provide a feeling of well being. However, to guarantee a safe, comfortable working environment, occupants should have control over the quantity and quality of light where visual tasks are performed. A mixture of lowered levels of background light from low-energy, ceiling-mounted fittings and daylight from windows, together with task lighting for close work, is often found to provide the most acceptable visual environment.

In achieving an acceptable comfort level of daylight, one of the main 'discomforts' to be recognised and resolved is the problem associated with glare (both direct and indirect). Treatment of this problem reflects a lighting strategy and has implications on the energy performance of the building. Glare is a function of contrast and brightness, which results from one of two causes. In the first, a bright light source (such as a sunlit window or a bright lamp) is viewed from a surrounding area that is in relative darkness. In this

Fig. 75   Light shelf designs (Source: Lawrence Berkeley National Laboratory, 1996)

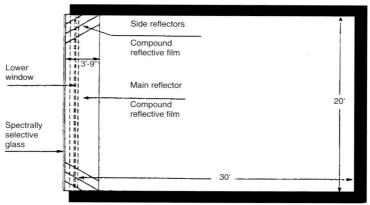

Side reflectors

Compound reflective film

Lower window

3'-9"

Main reflector

Compound reflective film

Spectrally selective glass

20'

30'

Floor plan of light shelf designs

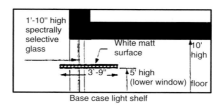

1'-10" high spectrally selective glass

White matt surface

10' high

3'-9"

5' high (lower window)

floor

Base case light shelf

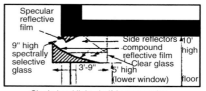

Specular reflective film

9" high spectrally selective glass

Side reflectors
compound reflective film
Clear glass

10' high

3'-9"

5' high (lower window)

floor

Single level light shelf (same section with and without reflectors)

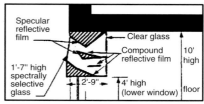

Specular reflective film

Clear glass

Compound reflective film

1'-7" high spectrally selective glass

10' high

2'-9"

4' high (lower window)

floor

Bi-level light shelf

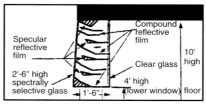

Specular reflective film

Compound reflective film

Clear glass

2'-6" high spectrally selective glass

10' high

1'-6"

4' high (lower window)

floor

Multi-level light shelf

case, glare results from excessive contrast and can be relieved by increasing the brightness of the surroundings. This type of glare causes discomfort to occupants, resulting in poor visual performance and potential dissatisfaction with the visual environment. Often complaints manifest themselves as more general dissatisfaction with the environment as a whole due to the subjective nature of the problem, which is not always obvious.

If a space is 'over-lit' by a source that is so excessively bright that the eye mechanism becomes saturated, the result is 'disability glare'. This, although less likely to occur, can be debilitating or even dangerous. For example, a window in sunlight at the end of a dimly lit corridor can suddenly plunge a pedestrian into relative darkness and could be hazardous.

Glare can be avoided by careful consideration of window design in relation to the room depth and height, surface attributes of the space and the relationship between the window, the exterior and the occupants. A room three metres high and six metres deep daylit on one side only should achieve a daylight factor of around 1.5–2 percent at the back of the room, for around 15–20 percent glazing/ external wall ratios. In terms of the visual environment created by this scenario, these levels are described as 'cheerfully daylit'. A height-to-depth ratio of 1:2 allows good light penetration for the aforementioned glazing ratios, which can be applied to rooms with higher and lower ceilings or deeper and shallower plans pro rata. The effect of this ratio is to limit the depth of a non-residential building to about 12 metres, assuming it is lit on both sides.

Glare problems can be solved by improving or reducing contrasts – for example, by increasing internal surface reflectance. In schools, glare often occurs when light-coloured walls are concealed behind posters or friezes; this could be alleviated by concentrating such material on walls adjacent to the window to allow the wall opposite the window to remain as free as possible of artwork and other coverings.

The urban building's floor layout and shape, besides responding to commercial considerations, should take into account the local users, modalities and cultural patterns of working, privacy and community, all of which have developed in relation to the locality's climate. This should be reflected in the building's floor-plate configuration, its floor-depth, the positioning of its entrances and exits, the provision for human movement through and between spaces, its orientation and its external views. The floor-plate configuration should provide a habitable environment, interest and scale, internally available sunlight, sufficient ventilation and so forth. For example, for offices, the workstations should not be located in the centre of each floor plate with the partitioned offices at the periphery, but should be reversed by planning the internal layout to enable the greatest number of users to receive natural sunlight.

Artificial lighting accounts for about 10 percent of the energy used in the typical large building (whereas carbon dioxide emissions are about 25 percent of emissions). This is the next area for design efforts to minimise energy consumption in the skyscraper's and other intensive building types' operational systems (see Fig. 76). Although artificial lighting loads can be reduced in part by making the building configuration maximise the accessibility of natural sunlight, they can also be reduced by using high-efficiency lamps, low-energy artificial lamps, electronic ballasts and high-quality fittings. For instance, replacing a 75-watt incandescent bulb with an 18-watt compact fluorescent bulb will, over the lifetime of the bulb, avoid emitting the equivalent of 4,300 Kg of carbon dioxide and about 10 Kg of sulphur dioxide from a typical generating plant (in the USA), which also creates acid rain (MacKenzie, 1997, p. 19, and Zeiher, 1996, p. 107; see Fig. 77). Energy savings could be achieved by providing lighting switching systems coupled with the building's automation systems (BAS), or using local controls and ambient light sensors to adjust artificial lighting, lending to the amount of natural light entering the building.

Artificial-lighting energy consumption can be improved in existing light fittings (either two or three tubes) by replacing these with

Fig. 76   Annual energy and peak demand savings for different building types (Source: Guzowski et al., 1994)

Building Use	Retail	School	Warehouse	Grocery	Manufact.	Office
Daylight strategy	Side and Toplight	Side and Toplighting	Toplighting	Toplighting	Toplighting	Sidelight Light shelf
Lighting load Equipment load Electric lighting system	21.5 W/m² 5.3 W/m² HID and	21.5 W/m² 2.6 W/m² Fluorescent	8.7 W/m² 3.2 W/m² Fluorescent	24.2 W/m² 10.7 W/m² Fluorescent	13.8 W/m² 10.7 W/m² Fluorescent	21.5 W/m² 5.3 W/m² Fluorescent
Lux level Electric control system	540 lux On/Off	540 lux Dimming to 37%	320 lux On/Off	800 lux On/Off	540 lux On/Off	540 lux On/Off Switching 2 of 3 lamps
Annual electric Energy savings	97500 kWh/yr (20%)	36,000 kWh/yr (19%)	227700 kWh/yr (28%)	190,00 kWh/yr (29%)	167000 kWh/yr (29%)	650,000 kWh/yr (51%)
Total reduction in lighting energy	92,410 kWh/yr (39%)	39,859 kWh/yr (35%)	298,000 kWh/yr (70%)	182,000 kWh/yr (35%)	161,900 kWh/yr (62%)	501,341 kWh/yr (60%)
Total reduction in peak load	23.4 kW (12%)	23.4 kW (123.5%)	95.8 kW Summer Decrease (19%)	5 kW (3%)	20 kW (6%)	123 kW (42.5%)

Incandescents	Compact fluorescents
25 watts	7 watts
40 watts	11 watts
60 watts	15 watts
75 watts	20 watts
100 watts	27 watts

Fig. 77 Comparable bulb outputs (Source: Goldbeck, 1997)

single high-output tubes with electronic ballasts and high-performance reflectors (e.g., "3M Silverlux") with a reflectivity of 0.8. This will increase the light output by about 60 percent compared to the old fitting. By using electronic ballasts, the lamp flicker common with most standard magnetic ballasts is eliminated. An additional gain is the instant start-up of electronic ballasts as compared to the longer inconvenient start-up time of most magnetic ballasts.

Transitional Spaces and Skycourts

The skycourt is essentially a recessed balcony area with full-height glazed doors which open out from the internal areas to the terrace spaces.

As an alternative to shading devices, the building can have cut-in transitional spaces at its facade to provide useable 'places-in-the-sky' as well as recessed shading (see Fig. 77). Here, a useful device is the use of recessed terraces or 'skycourts' to serve as interstitial zones between the inside and the outside. These parks-in-the-sky balance the inorganic mass of the building's hardware and components with an organic mass to effect a more balanced ecosystem.

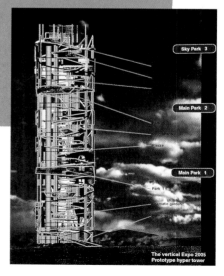

The vertical Expo 2005 Prototype hyper tower

Besides providing shading to that portion of the building, these skycourts can also serve multiple functions – as emergency evacuation zones, as areas for planting and landscaping, as flexible interstitial-zones for future expansion (e.g. in the event of future increase in permissible plot ratio), as areas for the future spatial addition of facilities (executive wash-

Fig. 78   Zoning for transitional spaces (Source:
Yeang, K., 'Bioclimatic Skyscrapers', Ellipsis, 1994)

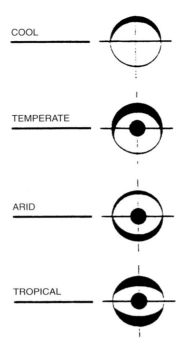

COOL

TEMPERATE

ARID

TROPICAL

ZONING FOR TRANSITIONAL SPACES

Diagram indicates the traditional placement of lobbies, stairs, hallways, and other annexed areas. These areas do not require total climatic control and natural venti- lation is usually sufficient. In tropical and arid zones, the transitional spaces are often located on the north and south sides of a building, where the sun doesn't penetrate too far into the rooms. Atria can also be used as transitional shaded spaces. In temperate and cold zones, the transitional spaces are located on the north side of the building, acting as buffer zones.

rooms, kitchenettes, etc.). The skycourts also give the skyscraper's users a more humane environment, letting them step out from the enclosed floor areas to directly experience the external environ- ment and to enjoy views.

These transitional spaces can also be located to protect the 'hot' sides of the urban building or to frame an important view. Position- ed either centrally or peripherally, such multi-storey transitional spaces essentially perform the same traditional transitional role (i.e. as in-between spaces) as the veranda in vernacular architec- ture. Such spaces are in effect 'open-to-the-sky' spaces under semi- enclosed conditions.

These spaces need not be totally covered from above. They might be shielded by a louvred roof to encourage wind flow to the inner areas while letting the hot air out. These spaces may even extend over the entire face of the building to create a multi-storey recessed atrium space, which might also serve as a wind scoop to vent the inner parts of the building. The hot air from the stack effect generates air movement through these atriums and is often used in hot-arid zones and in temperate zones (where the differences in the external and internal temperatures are sufficient to make this work).

The 'Fifth Facade' in the Building Shell: The Roofscape

The building's roofscape should be considered as the building's fifth facade. The roof of a skyscraper is less important thermally as compared to the roof of a lower-rise building type because of its small surface area compared to the extensive external-wall area. Furthermore, the skyscraper's extensive building height makes any roof overhangs irrelevant except for the uppermost few floors.

However, the direct solar-heat absorption of the roof in the last floors needs to be considered. In any case, much of the roof is usually occupied by mechanical equipment, which offers some insulation. The alternative is to have a roof canopy or pergola, or to provide a roof garden or permaculture.

In hot, humid climate zones, roofs should be constructed of low-thermal-capacity materials with reflective outside surfaces (where these are not shaded). The roof should preferably be of double construction and provided with a reflective upper surface.

The thermal forces acting on the outside of the tall building are a combination of radiation and convection impacts, among others. The radiation component consists of incident solar radiation and of radiant heat exchanged with the surroundings. The convective heat impact is a function of exchange with the internal air and may be accelerated by air movement.

Roofs and terrace areas in our building might also be vegetated. If buildings are designed with vegetated roofs, rainwater is retained and evaporated, new wildlife habitats are created, internal insulation is improved and energy consumption reduced. Roof vegetation can halve the rain water collected (which must be discharged), reduce heat-island phenomena in dense urban centres by cooling the air by evaporation, add thermal and acoustical insulation, protect and increase the lifespan of the roof underlay, recuperate habitable space for flora and fauna and increase the biodiversity of the locality.

The beneficial effects of vegetation are further treated in 'Passive Design by Vertical Landscaping'.

Fig. 79   Temperature of
interior surfaces of west
wall (Lat. 3°N, Haifa) (12 cm
concrete wall) (Source: Givoni, B., 1994)

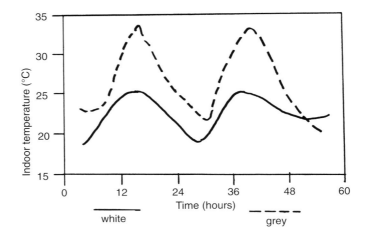

white   grey

Passive Design by Colour of the Building Envelope

Lighter-coloured materials, especially for the roof (which receives much more sun in summer, when it is not wanted, than in winter, when it is) can cut peak cooling needs by as much as 40 percent (see Fig. 79). Similar improvements can be achieved by placing vegetation, including large shade trees, around buildings (Flavin and Lenssen, 1994, p. 227). Both methods also contribute to reducing energy demand by minimising the 'urban heat island' – the boosting of urban temperatures when pavement and building surfaces absorb the sun's rays instead of reflecting them. Cooling needs can be cut 30 percent if enough trees are planted (Lawrence Berkeley National Laboratory, 1993).

The green skycourt

An effective protection against radiation impact for the external wall is the selective absorption and emission characteristics of certain materials, especially under hot conditions. Materials that reflect rather than absorb radiation, and which release the absorbed heat as thermal radiation more readily, bring about lower temperatures within the building. The external facade should be as light coloured as possible to reduce the heat-island effect and to lighten overall air-conditioning loads (Rosenfield et al., 1997). Special coatings are available to improve thermal performance of the base materials, and these also should be considered. Dark colours may be used on internal walls on those facades where high mass is used as part of the design strategy.

237

Passive Design by Vertical Landscaping

A crucial factor in ecological design is the counterbalancing of the inorganic characteristic of buildings with the organic or the biotic components analogous to the ecosystem. This can be achieved in the high-rise or large building by the incorporation of vertical landscaping into the built form (Rosenfield, op. cit.), demonstrated in figure 80. Roofs can be clad in turf or other vegetation, walls can have climbing plants, car-parking bays can have reinforced grass, roads can be lightly vegetated so that they are porous to dust and water.

In addition to its organic counterbalancing role, vertical landscaping also serves as part of the passive means to lowering the ambient temperatures around buildings, as well as the reduction of the overall urban heat island effect. Plants evaporate water through the metabolic process of evapotranspiration. The water is carried from the soil through the plant and evaporated from the leaves as a part of the photosynthesis process. The transpiration of water by plants helps to control and regulate humidity and temperature. Studies have shown that vertical plant cover on exposed wall surfaces improves the energy efficiency of the wall by up to 8 percent, partly by the pockets of air trapped and partly by preventing rain from filling the air voids in the building facade with water (i.e. a wet wall is a poorer insulator than a dry one).

Although soft-landscaping strategies for low- and medium-rise buildings are relatively well developed (common solutions include planter boxes and roof planting), they are notably less advanced for skyscrapers. Planting and organic material should be added to the facade, skycourts and balconies or in the inner skycourts of the urban building.

Vegetated roofs discharge less rainwater because:
- roof water is retained by plants and used in metabolic processes;

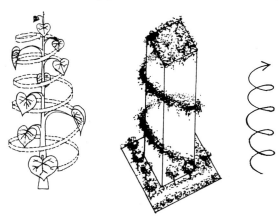

Fig. 80   Vertical landscaping (Source: Drawing adapted from Papanek, 1995)

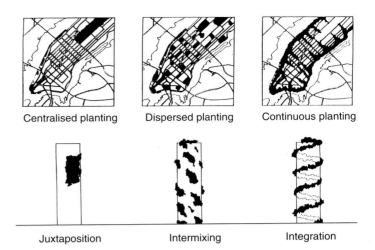

Fig. 81   Centralised planting and juxtaposition (Source: Yeang)

Fig. 82   Dispersed planting and intermixing (Source: Yeang)

Fig. 83   Continuous planting and integration (Source: Yeang)

Centralised planting   Dispersed planting   Continuous planting

Juxtaposition   Intermixing   Integration

⦿ flat roofs and terraces have additional retention capacity;
⦿ roof water is returned to the atmosphere by evaporation.
Roof surfaces and skycourts in the sky-
scraper have higher evaporation rates than ground surfaces
because they are more exposed to sun and wind. Vegetation
increases the area from which this evaporation can take place.
Cumulatively, these effects are significant. From a light rain shower,
there will be no runoff. From the moderate shower, only about half
of the precipitation will be discharged from the roof or skycourts.
With a heavy shower, there may be a delay of an hour or more
before there is any discharge from a previously dry roof. The total
annual discharge may be halved. This further justifies the provision
of skycourts in the skyscraper.

In any ecosystem, the climate is held to be the predominant
influence, even though other biotic factors such as flora, fauna and
soils have an effect on the system. In most urban locations, all that
remains of the site's original ecological components is probably the
topsoil and upper geological strata, and a much simplified and
reduced fauna. In any new built system, we should recognise the
fundamental ecological value of increasing the site's biological
diversity. Vertical landscaping in the skyscraper introduces organic
matter into an otherwise high concentration of inorganic mass on a
small site. It is contended here that there are essentially three basic
strategies for adding vegetation to buildings: juxtaposition, intermixing and integration (see Figs. 81, 82 and 83).

Plants can have aesthetic, ecological and energy conservation
benefits in addition to providing effective responses to wind and
rain. Planting can shade the internal spaces and the external walls
and can minimise external heat reflection and glare into the build-

ing. Plant evapotranspiration processes can be effective cooling devices on the facade, affecting the facade's microclimate by generation of oxygen and the absorption of carbon dioxide. For example, studies have shown that 150 m² of 'plant surface area' can produce enough oxygen for 1 person for 24 hours.

Vertical landscaping can also be used to reduce the ambient air temperature. Facade planting can lower ambient temperatures (in summer in temperate climates) at street level by as much as 5 °C (Hough, 1995). Heat loss in winter can be reduced by as much as 30 percent (op. cit.). Vegetation at the facade of buildings will obstruct, absorb and reflect a high percentage of solar radiation; the rest passes through the vegetation and reaches the building's surfaces. Plant leaves can be 1 °C lower than the ambient temperature, and damp surfaces like grass, soil or concrete can be 2 °C or more below and can contribute significantly to a cooler and healthier building (aesthetically), not to mention reducing energy costs accrued by air conditioning.

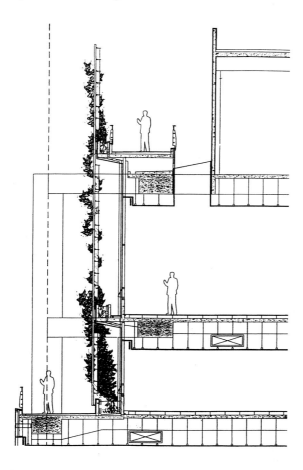

Fig. 84 **Building in section showing placement of planting**

Plant	Form-aldehyde	Benzol	Trichloro-ethylene
Banana	89	–	–
Bowstring hemp	–	53	13
Chrysanthemum	61	54	41
Dracoena deremensis (Janet-Craig)	–	78	18
Dracoena deremensis (Warneckii)	50	70	20
Dracoena deremensis (massangeana)	70	–	13
Dracoena deremensis (yellow-variegated)	–	79	13
True aloe	90	–	–
Ivy	–	90	11
Devil's ivy	67	73	9
Spathe flower	–	80	23
Creeping hairy spurge	67	–	–
Ficus benjamina	–	–	11
Gerbera	50	68	35
Green lily	86	81	–
Chinese evergreen (Aglaonema)	–	48	–
Philodendron (domesticum)	86	–	–
Philodendron (oxycardium)	71	–	–
Philodendron (selloum)	76	–	–

Biologically, the leaf is an efficient solar collector. In summer, the leaves take advantage of solar radiation, permitting air to circulate between the plant and the building; cooling takes place by means of a 'chimney effect' and through transpiration. In winter, the overlapping leaves form an insulating layer of stationary air around the building. Even in regions too cold for evergreens to grow, summer cooling may still be an important factor, lending an energy saving and biological validity to planting.

The tall building's facade area can be up to four or five times the area of the site, perhaps even more. If the facade is covered with planting in its entirety, the increase in vegetative cooling can be significant (see Figs. 83 and 84). The complete covering of the facade can contribute significantly to reducing urban heat-island effects. Externally, vegetation can also lower urban temperatures in the boundary layer by 1 °C, while vegetative canopies (i.e. trees) may lower external ambient temperatures by an additional 2 °C in the area under the canopy (Kurn et al., 1994).

Studies have shown that not only do plants process carbon dioxide and release oxygen into the air, they also remove formaldehyde, benzene and airborne microbes, thereby contributing to a healthier internal environment (the required concentration is one plant per square metre of internal space). The Boston fern removes 90 percent of the chemicals that cause allergic reactions (Wolverton, B.C., 1996) (see Fig. 85).

Vegetation on rooftops functions in the same way as at ground level in terms of its role in climate control. It would be a mistake to

consider the roof of a skyscraper too inhospitable an environment for climatically significant amounts of plant life. Hardy plants
can adapt to such environments with minimum soil depth or
humus content. Certain plants can grow on only seven centimetres
of soil consisting of pea gravel and silt sand. The depth of soil
needed depends on the type of plant. For example: grasses need
150–300 mm; ground cover and vines need 300 mm; low and medium shrubs, 600–750 mm; large shrubs and trees, 600–1050 mm.
New landscape at the roof or skycourt level can contribute to the
climatic conditions of the city by reducing heat absorption. Roof
gardens can also be used for urban agriculture (most vegetables
need no more than 20 cm of soil).

The humidity of vegetated areas is related to evapotranspiration. The evaporation effect depends on the albedo, morphology,
rugosity and articular resistance of the leaf surface (Mascaro, et al.,
1998). The relative humidity of the air (in a humid subtropical climate) under the vegetation is found to be between 3 percent and 10
percent greater than in the areas without vegetation. The bigger differences are in the summer (in the subtropical climate) because
this effect is proportional to the density of the vegetation leaves.
The smaller values are registered in the spring due to the action of
the winds and the existence of empty spaces in the crown during
the flowering period. Through rooftop vegetation, the toxins in the
rain are absorbed or destroyed in the soil, and the storm water is
retained for slow discharge. Evapotranspiration can be enhanced in
hot weather with nighttime spraying, which cools the building
below and reduces the need for air-conditioning. In temperate
zones in winter, the snow cover and the ice-dried fibre-matrix can
serve as insulation.

A single large tree can transpire 450 litres of water a day (equivalent to 960,000 KJ of energy in evaporation), which is rendered unavailable to heat surfaces. The mechanical equivalent would be to
have 5 average-size room air conditioners, each operating at 10,500
KJ per hour, running for 19 hours a day. Air conditioners only shift
heat from indoors to outdoors, and also use electric power. The heat
is therefore still free to increase urban air temperatures and therefore the heat-island effect (demonstrating, incidentally, the need for
designers to think in terms of the interconnectivity of environments).
But with the tree, transpiration renders this same heat unavailable.
To illustrate the importance of landscaping as a passive-mode strategy, one study of moisture sources in the urban air concluded that
advection and evapotranspiration are significant contributors to
humidity (after Kurn et al., 1994), and that a vegetated surface can
be as effective as a high-albedo surface in reducing the sensible

Fig. 86 Vegetation integrated into a design scheme

heat gain of urban air. Studies have shown that near-surface air temperatures over vegetated areas appear to be 1–2.25 °C lower than background air temperatures, and that vegetation may lower the urban temperatures in the boundary layer by about 1°–1.25 °C.

Ecologically, it is crucial to ensure physical contiguity between planting for encouraging species migration and contributing to greater diversity (see Fig. 86). To achieve physical contiguity in 'vertical landscaping' in the urban building, the system should be linked (e.g. using stepped planter boxes organised as 'continuous planting zones' up the face of the building). These would permit some extent of species interaction and migration, and would provide a link to the ecosystem at the ground level. The alternative option would be to separate the planting into unconnected boxes; however, this can lead to species homogeneity, which necessitates greater external inputs (regular human maintenance) to remain ecologically stable.

As a general design strategy, we should reintroduce as much as possible the indigenous vegetation of the locality. We will likely find

243

that the indigenous vegetation, reintroduced correctly within its own climatic range, typically requires little maintenance and makes low demands on other scarce resources such as water, fertilizer and energy.

In considering in totality the carbon dioxide emissions in the skyscraper and other intensive building types, one strategy that has been suggested to provide compensation to the embodied ecological impact of the production of the building arising from its CO and CO_2 emissions is to 'adopt' an area of rainforest equal to the extent of carbon dioxide emissions in the building. However, there is a flaw in this argument, in that ecological design is then regarded as a battle to reduce the elimination of rainforests by the timber-producing countries. While this is a goal that must be achieved, the designers should also contribute to the enhancement of the urban environment by adding more organic matter to the site and to the building equivalent to 1 m² of rainforest (capable of absorbing 1 Kg of carbon dioxide per year). It has been estimated that 200 trees are required to absorb the carbon dioxide emissions of one car (Zeiher, 1996). Therefore, to compensate for the car owners using the skyscraper, the equivalent in trees needs to be planted on the site or at that urban locality.

As a general strategy, the designer should free as much of the ground plane as possible and make available opportunities in the upper parts of the skyscraper for the colonisation of the building by flora and fauna.

Passive Design in the Use of Wind and Natural Ventilation

One of the key ambient sources of energy of a location is wind. By optimising the location's wind conditions during the time of the year and time of the day when good ventilation or when wind-assisted comfort-ventilation is needed, we can shape the urban building's floor plate and external wall for natural ventilation and more effective cooling.

Natural ventilation includes a number of ways in which external air and wind can be used to benefit the occupants of buildings. At the simplest level, natural ventilation ensures a fresh air supply to the interiors (e.g. through vents in double-skin facades or as simple vents installed above windows). However, it must be taken into account that this could create dustier or noisier internal conditions, especially at the lower levels of the high-rise (e.g. c. 5 to 8 storeys and below). Large projects may be dependent on the 'stack effect' in mixed-mode systems, whereby fresh air is allowed to enter at a low level, where it can be further cooled as it comes into contact with the thermal mass of the cold concrete floor slabs. As the air warms, it rises and is eventually expelled at roof level.

Properly designed natural ventilation solutions can result in savings in both capital costs and energy. In addition, it is also desirable to minimise the requirement for mechanical ventilation and air-conditioning systems in order to ensure a 'healthy' building. The use of natural ventilation is overwhelmingly supported by the fact that energy consumption is typically around half that of air-conitioned buildings in cases where natural methods are used. In addition, maintenance is reduced, there are fewer incidents of 'sick building syndrome' and there is a reduction in carbon dioxide emissions.

Enclosed central courtyards or atriums can be used to save energy by using the space as a means to bring fresh air into the building and to provide natural 'pre-heat'. In addition, a design incorporating such an atrium space should lend itself to natural ventilation by virtue of the fact that the inclusion of the atrium will modify the building form to one avoiding deep-plan accommodation in favour of a layout with windows on inner and outer facades, allowing for good cross ventilation.

Natural ventilation should be employed whenever possible without incurring heating (or cooling) energy penalties. Ventilation helps to control indoor air pollution by diluting stale indoor air with fresh outside air. Air movement will be encouraged by temperature and pressure differentials between inside and outside, particularly where temperature differences are enhanced by climate-sensitive considerations such as passive solar gains, atrium spaces and glazed courtyards.

However, not all buildings lend themselves to a completely natural approach; indeed, in winter, care must be taken to avoid over-ventilation and consequent energy penalties due to excessive fresh-air cooling. As a result, in larger buildings, 'mixed-mode' and 'displacement' ventilation systems have begun to emerge as a means to conserve energy in winter.

Fig. 87   Indoor air quality
factors (Source: HBI, 1994)

Contaminants	%	Building system problems	
Fungi	32	Ventilation inadequate	50.4
Dust	22	Filtration inadequate	55.6
Low relative humidity	16	Hygiene inadequate	41.7
Bacteria	13		
Formaldehyde	8		
Fibrous glass	6		
Exhaust fumes	5		
VOCs	4		
ETS	3		
High relative humidity	2		
Ozone	1		

Indoor air quality by problem type

Problem type	Average
Poor ventilation	56.6
Interior contamination	6
Microbes	16.9
Tobacco smoke	1.3

The argument against natural ventilation is that there is an
increase in air and noise pollution in the interior of the building. It is
held that this defect also applies to many mechanical systems, in
which a misplaced fresh-air inlet can recycle dirty air into the interiors.
Factors influencing Indoor Air Quality (IAQ) (see Fig. 87) include
ventilation, humidity, lighting, contaminations, furnishings and colour schemes, maintenance, cleaning, use of the building, building
management and noise. The three primary sources of poor indoor
quality are as follows: hermetically sealed buildings and their synthetic furnishings, reduced ventilation and human bio-effluents.
However, we will find that more than 50 percent of poor indoor
quality is due to inadequate ventilation (Robertson, J., 1990), and
natural ventilation can contribute to greater internal comfort and
user well-being. Typical air rates are about 0.5 to 3.0 ac/hr. depending upon density of occupation (Baker, 1992). Healthy humidity
levels range between 35 and 65 percent. Typical recommended
values for air exchange per occupant range about 5 to 25 litres/
second/person. However, IAQ can be enhanced by improved air
changes (e.g more than 4 ac/hr) and in some cases, architects have
increased air changes up to 6 to 8 ac/hr (Croxton and Childs in
Zeiher, 1996, p. 38). However, at this air-change level, and above,
there may be disruptions to work surfaces (such as displacement
of paper).

Smoke:

Annual median	80 µg/m^3	

SO:

Annual median	120 µg/m^3	if smoke < 40 µg/m^3
	80 µg/m^3	if smoke > 40 µg/m^2

Smoke:

Annual median	130 µg/m^3	

SO:

Annual median	180 µg/m^3	if smoke < 60 µg/m^3
	130 µg/m^3	if smoke > 60 µg/m^2

Smoke:

Annual peak	250 µg/m^3	(98th percentile of daily values)

SO:

Annual peak	350 µg/m^3	if smoke < 150 µg/m^3
	250 µg/m^3	if smoke > 150 µg/m^2

NO:

Annual peak	200 µg/m^3	(98th percentile of hourly values)

Lead:

Annual mean	2 µg/m^3	

A draft directive on air pollution by ozone (CCM (92) 236 final) sets the following standards:

Ozone:

8 hour mean	110 µg/m^3
1 hour mean	360 µg/m^3 health protection warning
1 hour mean	200 µg/m^3 vegetation protection threshold

Current standards in the USA (ASHRAE Standard 62-1989) have the proviso that ambient air quality requirements are site-specific and not regionally specific (i.e., they refer to ambient air quality at the proposed point of fresh air intake; see figure 88). A building's fresh air intake should be located away from loading areas, building exhaust fans, cooling towers and other sources of contamination.

Indoor materials also introduce pollutants. Volatile organic compounds (VOCs) are carbon-based chemical solvents distilled from petroleum or petroleum byproducts. VOCs are often carcinogenic and, in quite small amounts, can cause or contribute to a wide range of serious human ailments. These range from birth defects and metabolic disorders to kidney and lung disease, memory loss and respiratory problems. VOCs and other deleterious chemicals are found in many of the most widely used construction and interior materials, and are included to enhance the performance of the product. Current standards for VOCs specify outgassing levels of around 0.5mg/cu. metre.

Examples of construction materials that contain VOCs are plywood made with isocyanurates, vinyl flooring made with polyvinyl chloride (PVC), many paints, glues and adhesives, and almost all maintenance and cleaning products that are petrochemical- or solvent-based. These materials (and others) contribute to the 'sick building syndrome'. The designer should seek to avoid introducing these contaminants and other toxic materials as much as possible, since human inhabitants are biotic components in the ecosystem. Every major material used should be researched, its chemical make-up analysed, and then chosen or rejected on its environmental and IAQ impact, besides the ecological impact of its production and distribution. For example, interiors should use 100 percent wool carpet woven to avoid the use of adhesives in the backing (e.g. use jute underlays instead), low toxic-emission particle boards, low-VOC paints and other finishes, and factory finishing of those products which have off-gassing.

Ecological design needs to take into account indoor air quality and its impact on building users, since humans are also a species in the ecosystem. Good indoor air quality requires the following design provisions:

● An IAQ management plan for the construction process (e.g., protection of the ventilation system equipment and pathways from contamination), and after completion of construction and prior to occupancy, provision of cleaning requirements for ventilation system components and pathways exposed to contamination during construction.

● Reduction of construction contaminants in the building prior to occupancy (e.g. dust, particulates, contaminants related to water infiltration and VOCs).

● Provision of a minimum of 65 percent filtration of the return-air side of the HVAC system components during construction and replacement of air filtration media prior to occupancy.

● Provision of a permanent air monitoring system that should have the capability to monitor supply and return air, and ambient air at the fresh air intake, for carbon monoxide, carbon dioxide, total volatile organic compounds (TVOCs) and particulates.

● Implementing a building material emissions testing programme, particularly for:

    adhesives
    sealants
    caulks
    wood preservatives
    wood finishes
    carpet

carpet padding
paint
gypsum board
ceiling tiles
ceiling panels
insulation (thermal, fire, and acoustic)
composite wood products
gaskets
glazing compounds
control joint filter
wall coverings
floor coverings
work surfaces
HVAC sealants
HVAC linings
flexible fabrics

● Precautions to discourage microbial growth meeting the required standards, including anti-microbial agents. Components that must meet current standards include: air filtres and humidifier pads, HVAC insulation, carpets, adhesives, fabrics, polymeric surfaces (vinyl, epoxy, rubber flooring, laminates), ceiling tile coatings, paints and other elements of the building's interior.

Strategies to reduce the effects of VOCs include using higher ventilation rates (achieved by increasing air change per hour), use of charcoal filters or the selective use of certain houseplants (Baker, 1992). In the context of ecological building, it would be ideal to achieve and maintain comfortable room temperatures and humidities and to remove contaminants through internal foliage alone, without having to resort to any technical or mechanical means. The transpiration of the plants (and hence, evaporative cooling), as well as their production of oxygen and the elimination of contaminants play important roles.

Plant Life and Air Quality

Designers should be aware of studies carried out on the elimination of contaminants. Formaldehyde, benzol, and trichloroethylene were studied in relation to various plants (Wolverton, 1996; see Fig. 89). It is notable that the elimination is initially rapid but slows down after a period of two hours. It is as yet unclear whether the elimination process reaches a saturation point, and whether it decreases considerably or even ceases altogether after several days, since all studies to date have been carried out on a 24-hour basis.

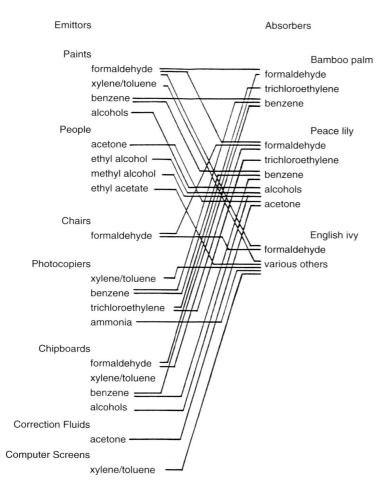

Emittors

Paints
- formaldehyde
- xylene/toluene
- benzene
- alcohols

People
- acetone
- ethyl alcohol
- methyl alcohol
- ethyl acetate

Chairs
- formaldehyde

Photocopiers
- xylene/toluene
- benzene
- trichloroethylene
- ammonia

Chipboards
- formaldehyde
- xylene/toluene
- benzene
- alcohols

Correction Fluids
- acetone

Computer Screens
- xylene/toluene

Absorbers

Bamboo palm
- formaldehyde
- trichloroethylene
- benzene

Peace lily
- formaldehyde
- trichloroethylene
- benzene
- alcohols
- acetone

English ivy
- formaldehyde
- various others

Fig. 89 Plant use for healthy environment (Source: Wolverton, B. C., 1996)

Long-term studies are needed to determine the actual consequences.

Certain plants are especially well suited to the elimination of contaminants. For instance, during a single day in an office, an ivy plant is able to eliminate 90 percent of the benzol contained in and released through tobacco smoke, artificial fibres, dyes and plastics. Aloe, bananas, spider plants and philodendron are effective agents against formaldehyde, which may seep from insulating foam and particle board. Trichloroethylene from lacquers and glues is best eliminated with the help of chrysanthemums and gerbera. It is certain that roots and the microbes symbiotic with root systems, and not the plants themselves, are largely responsible for eliminating contaminants.

With regard to humidification, plants are better agents than electrically powered air humidifiers or even humidifiers combined with air-conditioning systems, because they do not provide a favourable breeding ground for bacteria. Studies by Wolverton (1996) show that internal design using house plants can help provide an environment that mimics the way that nature cleans the earth's atmosphere. When their stomata open to absorb and release air and water, the internal air is stimulated (Madison, 1998). This allows the plants to capture toxins, which go to the root systems where microbes break them down. Saturation and re-release of toxins is not a problem; the removal rate actually improves with exposure. Plants also emit phytochemicals that suppress spores and bacteria (for example, the Boston fern). Another effective indoor plant is the gerbera daisy, which is held to be best at the removal of formaldehyde (found, for example, in facial tissues, carpets, gas stoves and plywood). The lady palm is best for removal of ammonia, and the peace lily is found to digest human bioeffluents. Wolverton (1996) considers the areca palm the more effective for indoor absorption. Wolverton (op. cit.) uses four criteria to rate the plants: removal of chemical vapours, ease of growth and maintenance, pest resistance and transpiration rate.

Plants can only create a definite decrease in temperature in summer when all surrounding surfaces, with the exception of windows, are intensely covered in internal foliage. Individual plants do not result in a noticeable change. Nevertheless, the use of plants in buildings should be given more attention in the future, since the overall effects are unquestionably positive, especially the psychological effect of verdant foliage on the occupant. However, it must be acknowledged that plants need an environment providing more than the minimum level for survival (i.e. the compensation point between photosynthesis and respiration). This can only be achieved by choosing the best plants and location and combining plants that have long life expectancies. Light being the source for photosynthesis, plants should especially be used in very bright, naturally lit areas such as winter gardens, atria and open glassed-in office areas. When lighting conditions are poor, the photosynthesis and pollutant breakdown of plants are equal. If forced to grow in an environment below the compensation point, they receive too little light, begin to fade after a while and eventually die.

Natural Ventilation

Natural ventilation may be used to increase comfort (air movement), for health (air change) or for building cooling (wind speed). The designer has to ascertain what should be the basis for the use of the wind forces for the locality in question.

There are two ways in which ventilation can improve comfort. One is a direct physio-logical effect; by letting in more wind by opening the windows (for example, through the use of wing-walls in combina-tion with adjustable shutters and spoilers), the indoor air speed is increased, which will make the occupants feel cooler. This approach is termed comfort ventilation. Introducing outdoor air with a given higher wind speed into a building may provide a direct physiological cooling effect even when the indoor air temperature is actual-ly elevated; this is particularly the case when the humidity is high, as the higher air speed increases the rate of sweat evapor-ation from the skin, thus minimising the discomfort that occupants feel when their skin is wet (e.g. with indoor air speeds of around 1–2 m/sec.). Ceiling fans operate on the same principle. The other method is an indirect one: to ventilate the building only at night, and to use this air to cool the interior mass of the building during the following day. During the day, the cooled air mass reduces the heat build-up rate of the indoor temperature and thus provides a cooling effect. This strategy is termed nocturnal ventilative cooling.

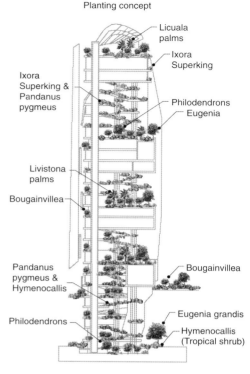

Planting concept

Licuala palms

Ixora Superking

Ixora Superking & Pandanus pygmeus

Philodendrons
Eugenia

Livistona palms

Bougainvillea

Pandanus pygmeus & Hymenocallis

Bougainvillea

Philodendrons

Eugenia grandis

Hymenocallis (Tropical shrub)

Fig. 90   Reintroduction of organic mass to urban site to counterbalance the inorganic nature of the site

In any event, the service cores of the green skyscraper should be naturally ventilated spaces and should have natural sunlight. At the same time, there should be good views to the outside wherever possible. This again reaffirms the preference for the peripheral-core position.

Further energy savings can be achieved through reduced requirements for mechanical ventilation, artificial lighting and the need for mechanical pressurisation ducts for fire protection. As mentioned earlier, the cores placed at the hot sides of the building can be used to absorb the heat build-up during the day and can then be flushed with cool air at night.

Internal open-plan layouts should ensure that occupants enjoy the advantages of daylight. By providing a view out from the lift-lobby areas, the skyscraper's users can experience the outside environment immediately as soon as they exit from the lift cars. As they step out from the enclosed elevator into the lift lobby, they receive natural sunlight and access to ventilation, and in this way, they enjoy a greater 'awareness of place'. In the central-core layout of the floor plate, the building's users exit the lift-car to enter into an artificially lit, 'location-unspecific' lobby, and often a dark and dingy passageway, which is far less pleasant by comparison.

Wind tunnel testing can be used in skyscraper design and for large buildings in the following applications: to assist in the design of the building's structural system; to aid the design of the skyscraper's outriggers (e.g., sky-bridges, steel balconies, etcetera); to provide the basis for the exterior facade design (the varying design wind speeds, surface pressures, suction effects and other factors); to identify opportunities for natural ventilation (at the lift lobbies, stairwells, toilets and other areas); to ascertain turbulences and conditions at the ground plane and in the skycourts (that is, to identify locations with uncomfortable or unsafe wind speeds; the Beaufort Scale No. 9 is equivalent to gale-force winds); and to ascertain opportunities for using wind generators.

Natural ventilation may be optional, and not a constant provision, in the low-energy large building. When needed (as when air-conditioning systems break down), and for natural cross-ventilation to be effective, the best window-opening arrangement is full wall openings on both the windward and leeward sides of the building in summer (or all the time in the hot-humid zones), with adjustable closure devices on the facade to assist in channeling the airflow in the required direction to match changes in wind direction.

Windows should be able to open to ideally large sizes when required (about 4 sq.m. for most lift lobbies and around 2 m² for staircases, achieving approximately 6 air changes per hour). However, high wind velocity at the upper reaches of the intensive urban buildings such as skyscrapers could make this impractical. Recessed windows with means of adjusting airflow and control of wind-swept rain could provide natural ventilation alternatives in the event of air-conditioning failure.

Air movement can generate cooling of occupants at air speeds of between 0.4 m and 3.0 m/sec. Air movement increases heat loss by both convection and evaporation; for instance, air movement of 1 m/sec. (walking pace) will reduce an air temperature of 30.25 °C to

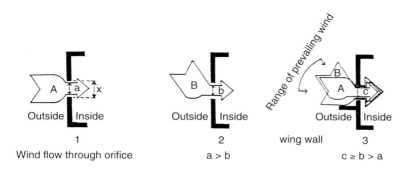

Fig. 91   Wind wing wall.
Diagrams 3, 4 and 5 illustrate the effect of a wing
wall on the facade of a
building (Source: Yeang)

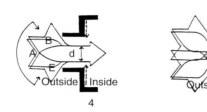

an effective temperature of 27.25 °C (Baker, 1992). Ceiling fans can be
very energy efficient when compared to air-conditioning, which
uses at least six times as much energy.

Wind performance grows exponentially as it moves upwards on
the skyscraper's facade. Therefore, if natural ventilation is used in
the building, then a series of modified venting devices for different
height zones is needed. The external facade can consist of a series
of systems (e.g., double-skin, flue wall, etc.) depending on the
desired thermal effect and venting system.

The taller the building, the greater should be its potential to
ventilate itself by the stack effect. In current apartment skyscrapers,
this phenomenon is counteracted by the mechanical pressurisation
of the corridor space and horizontal air barriers in high-rise buildings to dampen what would otherwise be excessive stack effect, at
the expense of using additional energy to ventilate the units.

Wing walls are also useful low-energy devices that can be used
in the skyscraper to capture wind using a 'fin' at the facade to chan-
nel wind into the insides of the building to increase the internal
airflow per hour to create comfortable internal conditions similar
to the effects of a ceiling fan (Givoni, 1994; see Fig. 91).

Natural ventilation is valuable for sustainable design as it relies
upon natural air movement and can save significant non-renew-
able-fuel-based energy by lowering the need for mechanical ventilation and air-conditioning. It addresses two basic needs in build-

ings: the removal of foul air and moisture, and the enhancement of personal thermal comfort (see Fig. 92).

Some European design codes dictate that in order to receive adequate natural light, the furthest desk from an outside wall should be in the range of 5 m to 7.5 m (e.g., depth at 2.5 times the external window height). It becomes impractical to attempt natural ventilation with greater depth of the floor plate, because natural ventilation becomes unsuitable – the air tends to become contaminated long before it is exhausted to the outside.

As mentioned earlier, in cities with higher levels of outdoor air pollution, natural ventilation to the useable areas becomes problematic. Air needs to be introduced into the space via air-conditioning systems with efficient filters to optimise the quality of the internal environment. External traffic noise might also militate against natural ventilation through open windows, unless suitable acoustical barriers exist. Skycourts might need sliding screens as wind breaks to protect from those instances of high wind speeds.

Natural ventilation, therefore, is suitable only for selective areas in the tall building such as the lift-lobbies, staircases and toilets, which can have openable windows or air gaps to the exterior, but also should be ventilated by a calculated percentage of air loss that is permitted to seep in from the air-conditioned spaces. Large balconies can have full-height adjustable sliding doors to serve as operable vents in cases where such natural ventilation is needed.

Fig. 92 User's most frequent internal environmental complaints (Source: Clinic for Occupational Medicine, Orebro, Sweden, 1994)

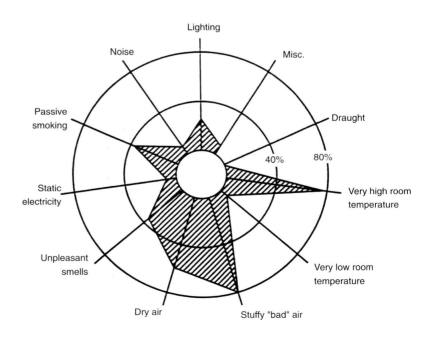

	Typical air-cond. office (kwh/sq.m.) (kwh/sq.m.)	Good practice open-plan office with nat. vent.
Heating and hot water	222	95
Lighting	67	32
Fans and pumps	61	5
Refrigeration	33	0
Catering	7	4
Total	390	136

Fig. 93   Comparison of energy use between an air-conditioned office and a naturally ventilated office (Source: BRECSU, Energy Consumption Guide 19)

In temperate and cold climates, the design challenge is to restrict incoming air to achieve the minimum necessary fresh-air change without causing cold draughts or excessive heat loss. Even under calm winter conditions, differences in temperature between the interior and outside air usually will create sufficient stack effect to draw in fresh air. The stack effect is brought about by warm air rising up to be exhausted through high-level outlets and so drawing in colder, heavier air from the outside. Natural ventilation is often not used in most user areas in tall buildings in cold and temperate climates, partly because of excessive wind speeds at higher levels but also due to problems arising from stack-effect. A common strategy in temperate climates is to use tempered or cooled minimal mechanical ventilation with natural ventilation (or fan-assisted ventilation), which varies depending upon the seasons, while limiting winter and summer design temperatures to 19 °C and 25 °C, respectively. Treating each season differently can enable the skyscraper to achieve acceptable energy consumption targets (around 150 Kw/hr/m² or less over a normal year). Natural ventilation for comfort and health purposes is frequently used for residential towers in warm-humid climates, where average wind speeds are lower (and often less frequent) than those in temperate latitudes and where stack effects are small. Most of today's passive solar and energy-efficient buildings are reliant on effective natural ventilation as one of their main strategies for the maintenance of thermal comfort (see Fig. 93).

Naturally-ventilated tall buildings should be shaped and oriented to maximise exposure to the required summer wind direction and designed with a relatively shallow plan (about 14 m external wall-to-wall floor-plate depth) to facilitate the flow of air through the building as cross-ventilation. Solar-heated buildings require particular attention in order to optimise both the solar and ventilation requirements. Ideally, solar orientation and breeze paths need to coincide.

To be effective for personal comfort, the air path through the building must pass through the zone frequented by the building occupants (that is, within about 2.0 m from floor level). Airflow above the heads of occupants in offices is of little value in summer (except for night cooling purposes), but can be useful in winter for achieving minimum ventilation needs while avoiding drafts.

Natural ventilation can be further induced by creating different pressures across the building (Aynsley, 1998). The essential principle is that a building's walls obstruct airflow, and so we need to create wind pressure differences between windward and leeward walls. The effective pressure difference tends to be about 1.4 times dynamic pressure at the projected floor-overhang level or at the location of any wing walls in the facade. Where wall openings are about 15–20 percent of wall area, the average wind speeds through wall openings can have the potential to be 18 percent higher than the local wind speed.

In temperate and cold climates during winter, it is necessary to ensure adequate exchange of indoor air to maintain indoor air quality. This natural ventilation can be achieved using wind pressure or the stack effect. During summer and in warm humid climates, natural ventilation with wind pressure and/or stack effects are useful for achieving air flow or indoor thermal comfort. The urban building in hot-arid desert climates can benefit from natural ventilation by using the stack effect to draw air through evaporative cooling systems or by wind pressure at night to enhance night cooling of the building. Natural ventilation during the day can be achieved through an evaporative cooling system (to ensure that the temperature of the incoming air is lowered and that its relative humidity is raised).

The ground floor of the tall building could be entirely open to the outside space and used as a naturally ventilating lobby wherever possible in hot-humid zones and in temperate zones (in the summer). It need not be enclosed nor air-conditioned, if it is to be effective as a 'transitional space' between the outside of the building (the street environment) and the building's lift lobby. Care, however, should be taken to keep out wind-swept rain and to avoid wind turbulence in these areas.

Passive Cooling Systems

Passive cooling applies to various simple cooling techniques which enable the indoor temperature of buildings to be lowered through the use of natural energy sources (Givoni, 1994).

Bioclimatic design techniques in hot-humid equatorial regions involve architectural designs and choices of materials that provide better comfort conditions (moderately better than the ambient internal conditions) while minimising the demand for energy used to cool a building. This involves minimising heat gain by the building, minimising solar heating of the building envelope and solar penetration through windows, providing comfort by natural ventilation and other techniques. The architectural means for achieving this have been discussed above (e.g., layout of building, orientation, the number, size, location and details of its windows, shading devices, the thermal resistance and heat capacity of its envelope). Appropriate applications of these design elements can bring the average indoor temperature to an improved comfort level better than the outdoor average. On the other hand, various passive cooling systems are capable of transferring heat from a building to various natural heat sinks by providing 'active' cooling through the use of passive processes, which often use heat-flow paths that do not exist in non-green buildings that lack these systems.

The appropriate architectural bioclimatic design for the region of the project site with hot summers can be considered as a precondition for the application of passive cooling systems, and the two approaches will supplement and reinforce one another. Buildings can be cooled by passive systems through the utilisation of several natural heat sinks such as the ambient air, the upper atmosphere, water and the under-surface soil. Each of these cooling sources can be utilised in various ways, resulting in different systems. The various passive cooling methods have been identified (Givoni, 1994) as the following:

● Comfort ventilation: providing direct human comfort, mainly during the daytime.

● Nocturnal ventilation cooling: cooling the structural mass of the building interior by ventilation during the night and closing the building during the daytime, thus lowering the indoor daytime temperature.

● Radiant cooling: transferring into the building cold energy generated during the night hours by radiant heat loss from the roof, or using a special radiator on the roof, with or without cold storage for the daytime.

● Direct evaporative cooling: mechanical or non-mechanical evaporative cooling of air. The humidified and cooled air is then introduced into the building.

● Indirect evaporative cooling: evaporative cooling of the roof, for example by roof ponds. The interior space is cooled without elevation of the humidity.

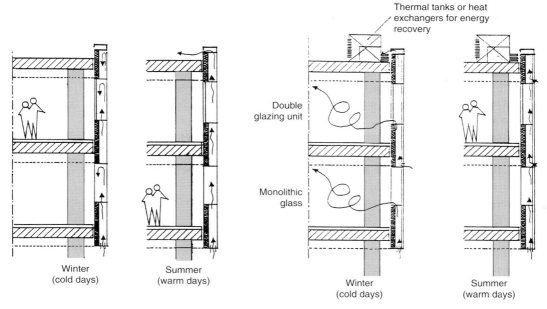

Thermal tanks or heat
exchangers for energy
recovery

Double
glazing unit

Monolithic
glass

| Winter
(cold days) | Summer
(warm days) | | Winter
(cold days) | Summer
(warm days) |

Mixed-mode systems   Passive systems

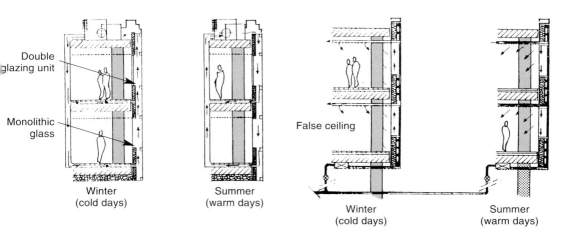

Double
glazing unit

Monolithic
glass

False ceiling

| Winter
(cold days) | Summer
(warm days) | | Winter
(cold days) | Summer
(warm days) |

Full-mode systems   Ventilation or pumping systems with
heat exchangers and/or pumps

**Fig. 94   Double facade
systems (Source: Perma-
steelisa)**

259

● Cooling of outdoor spaces: Cooling techniques that are applicable to outdoor spaces, such as patios, which are adjacent to a building.

Mixed-Mode Operational Systems

Having optimised all passive-mode strategies, the next stage is to investigate what mixed-mode systems could be used (see Fig. 94). In temperate zones, the basic mixed-mode approach encourages natural ventilation in summer and mid-season when outside temperatures are conducive (depending on internal gains and other factors). In winter, energy losses are minimised by changing over to a mechanically-assisted ventilation system, employing some re-circulation of extracted air or preferably heat recovery from the extraction system. Ventilation air is then supplied to the building, tempered to around 2 °C below the set-point temperature. Energy is saved by virtue of the fact that ingress of cold air via opened windows is virtually eliminated, as this could result in a need to heat the space further to compensate for excessive cold air infiltration (see Fig. 95).

In temperate climates, the bioclimatic strategy is to extend the mid-seasons (and as mentioned earlier, reduce the need for heating in winter and air-conditioning in summer through the use of a multiple-skin facade system that has different modes during the winter and summer). In winter, a part of the facade can become a trombe wall and absorb the heat from the sun. Part of the wall also lets in the eastern and western sun. In the evening, the trombe wall is used to heat up the interior and the internal blinds are pulled shut to insulate the insides to keep the heat in. The same wall is insulated on the outside in the summer to reduce the solar insolation and natural ventilation is used. In the summer nights, the mass of the trombe wall is ventilated to release the accumulated heat.

Displacement Ventilation

In this mixed-mode approach, generally, the system employed introduces air at the floor level, tempering the occupied zone (only) to around 1.8 m above floor height. There are three main benefits to this technique:
● Energy is saved by allowing the space above 1.8 m to 'float'. For this reason, a higher than normal ceiling height is generally required to gain maximum benefit (a floor-to-ceiling height in the order of 3 m is usual).

Fig. 95   Mixed-mode system (example) (Source: Yeang)

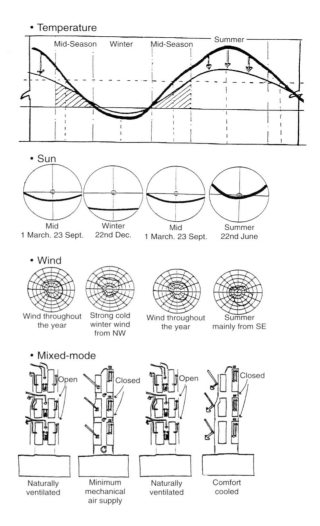

• Temperature

Mid-Season   Winter   Mid-Season   Summer

• Sun

Mid
1 March. 23 Sept.

Winter
22nd Dec.

Mid
1 March. 23 Sept.

Summer
22nd June

• Wind

Wind throughout
the year

Strong cold
winter wind
from NW

Wind throughout
the year

Summer
mainly from SE

• Mixed-mode

Open   Closed   Open   Closed

Naturally
ventilated

Minimum
mechanical
air supply

Naturally
ventilated

Comfort
cooled

● Contaminants are encouraged to rise towards a high-level exhaust zone, and incoming clean air travels upwards from a low level.

● In summer, if the space temperature rises above the design set-point of say 27 °C, air can be introduced at a low level at the outside air temperature (which will be slightly cooler). The upward movement of this cooler air should produce a cooling sensation and will improve comfort for the occupants.

In addition, the inert mass of concrete floors and ceiling voids may be used to provide some additional free cooling, as the slab temperature should be below the room air temperature. This can provide a further 1 °C or so of free cooling. The cooling effect can be enhanced by running cold outside air through the system overnight

Decision criteria for air-conditioning systems (e.g. VAV, FCU) vs natural ventilation systems				
Design criteria	Variable air volume	Fan-coil units	Displacement ventilation	Natural ventilation
Ease of installation	1	3	5	5
Commissioning requirements	3	3	5	5
Floor-to-floor height	2	3	3	2
Temperature control	4	5	2	1
Humidity control	2	3	4	1
Multi-zone control	5	5	5	1
Air movement	4	3	4	2
Air cleanliness	4	3	4	2
Odour control	1	2	4	2
Noise control	2	3	4	1
Flexibility	1	2	3	3
Capital cost of plant	3	2	4	5
Maintenance costs	3	2	4	5
Running costs	3	4	5	5
Total	37	43	56	41

A score between 1 and 5 is given with 5 representing a positive feature.

Fig. 96   Decision criteria for air-conditioning systems (e.g., VAV, FCU) vs natural ventilation systems (Source: Yeang)

for a few hours to remove residual slab heat. The effectiveness of this approach depends on the finished ceiling arrangement; for instance, suspended ceilings restrict the flow of air up and 'into' the slab.

Passive Solar Systems

Passive solar systems are essentially mixed-mode systems, as they are usually not full HVAC or fully-active systems. Like a living organism, the building using passive solar operational systems continuously seeks the path of the sun. The building becomes a skin that orients its occupants to a universal calendar. The hours of the day and the seasons of the year synchronise the comfort (and aesthetics) in the natural world with those inside. It is projected that in approximately 20 to 30 years, solar energy will be depended upon to provide an estimated 20 percent to 30 percent of the world's energy requirements (Daniels, 1995). Clearly, the sooner that measures are taken to convert existing skyscrapers and other large buildings to use and optimise on solar energy, the better (see Fig. 97).

Solar energy can be harnessed in many ways for the purposes of heating, cooling and lighting. The sun delivers approximately 5,000 times more energy per year than is consumed worldwide. Atmospheric solar radiation is approx. 1,300 W/m², of which approximately

1,000 W/m². reaches the Earth's surface (Daniels, 1995). The term 'passive solar' refers to systems that absorb, store and distribute the sun's energy without relying on mechanical devices like pumps and fans, which require additional energy. Passive solar design reduces the energy requirements of a building by meeting part or all of its daily heating, cooling, and lighting needs with solar energy.

The three most common passive solar design systems are direct gain, indirect gain, and isolated gain (see Fig. 97). A direct-gain system allows sunlight to pass through windows into an occupied space, where it is absorbed by the floors and walls. In an indirect-gain system, a medium for heat storage such as a wall located in one part of a building absorbs and stores solar heat. The heat is then transferred to the rest of the building through conduction, convection or radiation. In an isolated-gain system, solar energy is absorbed and stored in a separate area, such as a greenhouse or solarium, and then distributed to the living space through ducts. In order to conserve more energy, insulation is incorporated in the most effective passive solar designs.

In climatic regions with wide seasonal fluctuations in temperature, passive solar heating systems typically require some kind of backup heating. Nonetheless, passive systems contribute significantly to energy savings. Studies show that passive features typically add 5–10 percent to the cost of construction, yet the cost savings generally pay back the cost in 5–15 years, depending on geographical location and regional energy costs.

In active solar systems, solar collectors are used to convert the sun's energy into useful heat for hot water, space heating or industrial processes. Flat-plate collectors are typically used to gather the sun's energy. These are most often light-absorbing plates made of a dark-coloured material such as metal, rubber or plastic, which is then covered with glass. The plates transfer heat to air or water circulating above or below them, and the fluid is either used for immediate heating or stored for later use. These systems are referred to as 'active' because externally electro-mechanical powered equipment such as a fan or pump is used to move the fluid. In residential homes, active solar systems are generally designed to provide at least 40 percent of the building's heating needs.

Open-loop solar hot water systems are sometimes classified according to temperature. Mainly used for heating swimming pools, low-temperature unit collectors warm water up to 7 °C. Pool heating systems typically use pumps to circulate the water; storage tanks are generally not needed because the water is in constant circulation. In medium-temperature systems, the temperature rises to 7 °C to 28 °C for water and space heating in residential and commer-

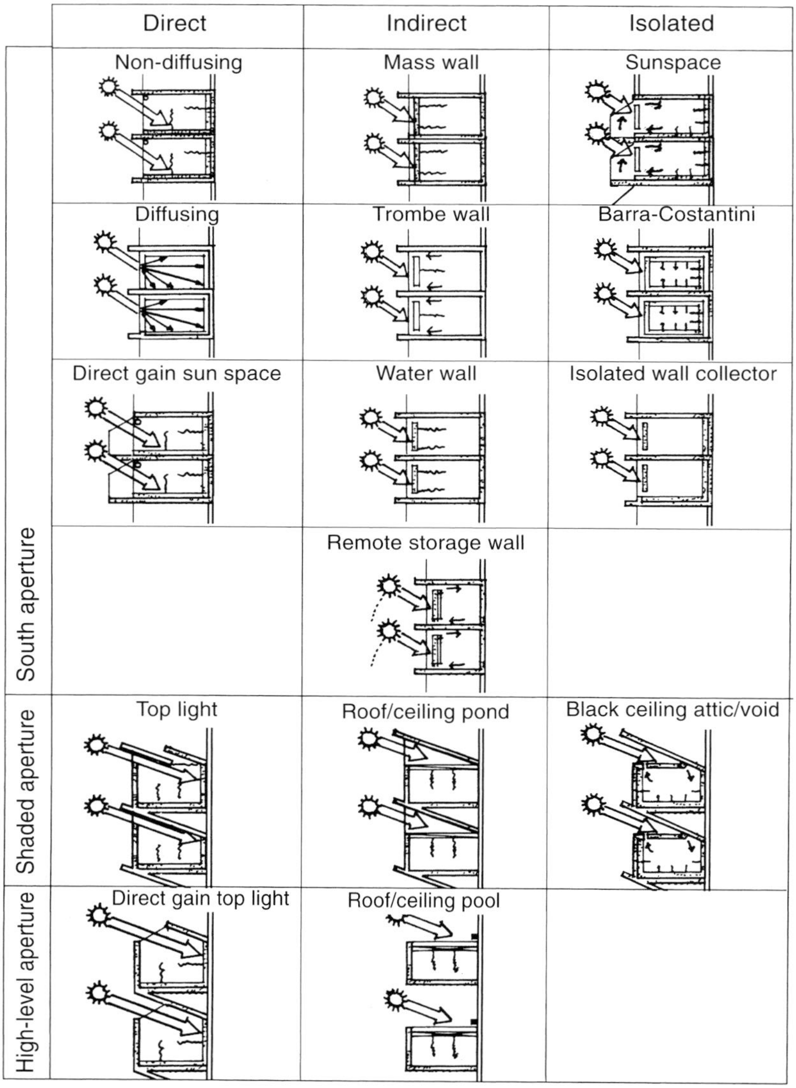

Fig. 97   Generic passive
mixed-mode solar systems
(adapted: Baker, 1992)

	Direct	Indirect	Isolated
South aperture	Non-diffusing	Mass wall	Sunspace
	Diffusing	Trombe wall	Barra-Costantini
	Direct gain sun space	Water wall	Isolated wall collector
		Remote storage wall	
Shaded aperture	Top light	Roof/ceiling pond	Black ceiling attic/void
High-level aperture	Direct gain top light	Roof/ceiling pool	

cial buildings and for industrial-process heat. In high-temperature collectors, water warmed to 28 °C or more is used for heat and hot water or for industrial processes such as cooking, washing, bleaching, anodizing and refining.

One common type of solar water heater, the thermosiphon system, uses collectors and circulating water, yet it is actually a passive system since no pumps are involved. In this type of system, the storage tank is installed above the collector. As water in the collector is heated and becomes less dense, it rises into the tank by convection and cooler water in the tank sinks into the collector.

Passive solar technology can also be used for cooling purposes. These systems function by either shielding buildings from direct heat gain or by transferring excess heat outdoors. Carefully designed elements such as overhangs, awnings, and eaves shade windows from the high-angle summer sun, while allowing the light from the low-angle winter sun to enter the building. The transfer of heat from the inside to the outside of a building may be achieved either through ventilation or conduction, in which heat is lost through a wall or floor. A radiant heat barrier, such as aluminium foil installed under a roof, is able to block up to 95 percent of the radiant heat transferred from the roof into the building. The Florida Solar Energy Center in Cape Canaveral has found that radiant heat barriers are the most cost-effective passive cooling option for hot southern climates.

Water evaporation is another effective method used to cool buildings, since water absorbs a large amount of heat from its surroundings when it changes state from a liquid to a gas. Fountains, sprays, and pools provide substantial cooling to the surrounding areas. The use of sprinkler systems to continually wet a building's roof in hot-dry zones during hot weather can reduce its cooling requirements by 25 percent. Transpiration, the release of water vapor through pores in the skins of vegetation, can reduce the air temperature in an immediate area by −15 °C to −10 °C.

Numerous approaches to active solar cooling have been developed as well. Provided sub-soil temperatures permit, one method includes the use of earth cooling tubes – long, buried pipes with one end open to the outside air and the other opening into a building interior. Fans draw hot outside air into the underground pipes, where the air loses heat to the soil, which remains at a relatively constant, cool temperature year-round. The soil-cooled air is then blown into the building and circulated. In active evaporative cooling systems, fans draw air through a damp medium, such as water spray or wetted pads. The evaporation of water from the medium cools the air stream.

Desiccant cooling systems are designed to both dehumidify and cool air. These systems are particularly well suited to hot, humid climates where air conditioning accounts for the majority of a building's energy requirements. Desiccant materials such as silica gels and certain salt compounds naturally absorb moisture from humid air. Moisture-laden desiccant materials will release the stored moisture when heated, a characteristic that allows them to be reused. In a solar desiccant system, the sun provides the heat needed to recharge the desiccants. Once the air has been dehumidified with desiccants, it can be chilled through evaporative cooling or other techniques to provide relatively cool, dry air.

The Full Mode (or Active) Operational Systems

Following the previous design considerations, the designer can then next coordinate the passive-mode systems, the mixed-mode (or background) systems with the full-mode (or active operational) systems to be provided (see Fig. 76). Only after all the passive-mode and mixed-mode systems have been maximised can design solutions be sought in operational and active systems (e.g., full lighting, heating and air-conditioning systems), depending on the level of internal environment servicing to be provided. As mentioned earlier, a significant extent of the energy used in the building is during its operation phase by internal M&E engineering systems. For example, in the 50-year life cycle costs of a typical commercial skyscraper (in Nikken Sekkei, 1996), the building's energy costs are 34 percent or more of the total costs.

In the operation of a building in the temperate zone, its space heating is at least 48 percent of its total operational energy use.

Fig. 97 indicates those areas where greatest energy savings can be achieved by passive-mode means (i.e. at a building's operational phase). The key economic benefit of a bioclimatic approach is a building with lowered capital and operating costs, due to lower energy consumption during its operational mode (savings ranging from 20 percent to 40 percent of energy costs are often achievable over the building's entire life cycle). Operational cost savings mean less use of electrical energy, particularly from non-renewable resources, which further reduces overall emissions of waste heat and particulates and the resultant urban heat-island effects. While this when translated into actual monetary costs is small by comparison to total costs, the greater benefit is in enhanced productivity (e.g. reduction in eye irritation, etc.).

In temperate and cold climates, the building should perform differently in relation to the changing seasons of the year of the locality. Significant energy savings can also be achieved by accepting a lower room design temperature in winter in temperate zones. In full-mode buildings the cooling load of a skyscraper or similar large buildings in their air-conditioning systems can be decreased by approximately 10 percent by increasing the designed room temperature by 1 °C.

Comparing three types of air-conditioning – variable air-volume system (VAV), fan-coil units, and displacement ventilation – we can make a simple design decision chart.

Significant energy savings can be achieved by using upgraded mechanical ventilation and air-conditioning systems (e.g., chillers, pumps, cooling towers and other systems). The green skyscraper and intensive building type should seek to achieve energy use of about 100 kWh/m² per year or less, compared with 230 kWh/m² per year for typically fully air-conditioned (and heated, if in a temperate zone) buildings and about 150–250 kWh/m² per year for typical un-air-conditioned offices.

The following design guidelines should be applied to buildings that are unavoidably fully air-conditioned:
- Avoid dual-duct systems or fixed-volume terminal reheating, as both involve cooling air and reheating it as required at the point of use.
- Avoid high-velocity systems as they consume more energy, but use less material.
- Use VAV systems (and FCUs) to provide local control.
- Employ heat reclaimed from parts of the building with a heat surplus.
- Avoid any system incorporating CFCs and HCFCs.
- Select equipment having maximum efficiency.

Air-conditioning system efficiencies can also be improved in a big building by using outgoing air to pre-dry the incoming air via a simple device with no moving parts, or to use heat exchanges three to ten times larger; making the chiller spin at just the right speed will contribute to 20 percent of the total energy savings. Another 40 percent of the energy savings would be in the big 'supply fan' that blows chilled air around the building, and the other 40 percent in the pumps and in the cooling tower fans that dissipate heat to the outdoors.

Water Conservation Strategies

Ecological design should diligently respect water resources. The United States, for example, uses 3 times as much water per day (578 litres per person) as the average European country, and considerably more than most developing countries (e.g. around 160 litres/per person/day) (Graves, 1998). It is important to recall that 90 percent of the global supply of water is in the form of salt water. Only 3 percent is fresh, and two-thirds of that is ice. Although about 4 trillion gallons of water falls globally daily in the form of precipitation, most of that disappears in evaporation and runoff. Population increases in urban centres, notably in developing countries, are exerting tremendous pressure on groundwater resources. Untreated human wastes remain the biggest pollution threat to water resource.

Groundwater is the main source of base floors for rivers, lakes and wetlands. It is also an effective buffer against drought. As global warming is expected to alter recharge patterns, the buffering action is going to become more important. When rainfall is insufficient and rivers run dry, groundwater remains a dependable source of water for drinking and irrigation. But worldwide groundwater supplies are being depleted at an alarming rate. More than 50 percent of the global water for drinking, washing and irrigation of crops comes from underground. As the demand for groundwater increases, many of its sources are polluted by the dumping of toxic wastes and by the slow seepage of agrochemicals through the soil. Overmining of groundwater in many crop-producing areas is lowering the water tables, and the change in flow paths of groundwater brings pollutants into non-contaminated areas.

Precipitation falling on the site should in theory eventually recharge aquifers and natural waterways without deteriorating their quality or quantity. Where contamination occurs, the building should provide for removal and/or recovery of contaminants. Water entering should not diminish its source, and should be returned to the natural environment in a state that enhances the aquatic habitat.

Installing water-saving fixtures is easy, inexpensive and provides immediate cost savings. The installation of new low-flush toilets, faucet aerators, efficient showerheads and efficient appliances can cut water use by over 30 percent. Numerous models of toilets, showerheads and faucet aerators provide a large array of features and design options.

Water conservation should be promoted by specifying low-use water fixtures and appliances and by water recycling systems that limit demand and reduce sewage. To be able to meet the demands of the future, immediate improvements are needed in techniques for conservation, collection, storage, treatment and reuse. Water conservation includes using water that is of lower quality than drinking water, like reclaimed wastewater effluent, gray water or runoff from ground surfaces, for such uses as toilet flushing or the irrigation of vegetation. Rainwater runoffs can be directed to

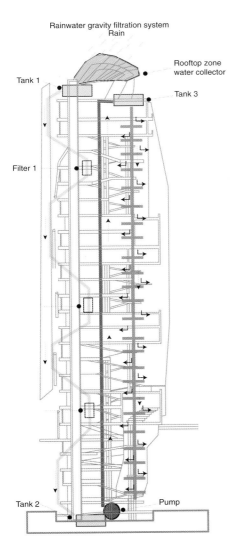

Fig. 98 **Rainwater purification system**

plants with high water demands, although drought-tolerant and xerophyte plants should be used where appropriate in landscaping.

The former standard toilet of 13 to 19 litres per flush is the largest water consumer in residences and in offices. To conserve water in new construction, the maximum permissible water use per flush is 6 litres (in new water-conserving toilets) according to many municipal building codes. Every toilet manufacturer now provides a number of models that meet this criterion. Before specifying, evaluate the operational noise, solids evacuation, bowl cleaning and water surface area. Double-flush units save water by providing a partial flush for liquid wastes and a complete 6-litre flush for solid waste. A waterless or composting toilet may be appropriate in some environments.

Urinals should have a maximum flow rate of 4 litres per minute (gpm) and be spring-loaded. Lavatory fixtures should also be spring-loaded and have a maximum flow rate of 8.3 litres/minute gpm at a test pressure of 0.042 kg/mm². Most water systems operate in the 0.018 to 0.028 kg/mm² range, but the higher-test pressure ensures that a conservation device functions over a wide range of pressures. Electronic control devices can be used in commercial installations, but local maintenance personnel should be knowledgeable in their repair.

Shower fixtures should be rated for a maximum flow rate of 9.5 litres per minute at 0.056 kg/mm². In public facilities where water use is more difficult to monitor, shower fixtures can have a timed cycle after they are activated by the user. They can also be spring-loaded with a chain operator, in which a hand chain-pull and spring automatically shuts off the fixture.

In kitchen and laundry areas, commercial appliances should be specified as water-saving models. Garbage disposals exert a huge load on wastewater treatment facilities and should be avoided. Composting provides a more useful alternative for food waste. The water used in typical fixtures is summarised below:

Fixture	Delivery rate in litres per minute
Kitchen faucets	9.5 or less
Bathroom faucets	9.5 or less
Showerheads	9.5 or less
Toilets and flush valves	6 or less per flush
Urinals and flush valves	6 or less per flush

We might direct our design efforts to reducing the rate of use. Potable water usage will be minimised through the use of:
- floor restrictors on most outlets
- low-flush toilets
- selection of landscape plant species that minimise the need for long-term irrigation
- connection to the block's non-potable recycled water reticulation system (e.g., connected toilets and external garden taps; see Fig. 99)

Biological Wastewater and Sewage Recycling Systems

Biological sewage treatment is an alternative to conventional sewage treatment, and biological wastewater systems have many environmental and economic advantages. Though they do not actually save water, they greatly contribute to the conservation of fresh water (e.g., Todd, N.J., and Todd, J., op. cit., p. xvii). These systems need to be evaluated in relation to the interactions framework of chapter 3 (as L12).

Biological wastewater systems should be a version of natural processes. Some systems link wetlands and marshes to purify water, while others treat waste in large solar greenhouses or use algal turf scrubber systems. In a marsh-type system, sewage water passes through a series of wetlands where it is purified by water-loving plants and microorganisms, emerging cleaner than Class 1 drinking water. This type of system is low-cost, low-maintenance and low-tech, yet it requires a great deal of land. The greenhouse or solar aquatics approach requires much less land. In this system, wastewater passes through a series of tanks and is gradually purified by plants, bacteria, invertebrates, fish and sunlight.

Biological systems use much less energy, capital and far fewer chemicals than regular waste treatment plants. They are less expensive to operate, can be an attractive educational feature in a building, and can provide natural habitat, fertilizer, and food. Biological systems can be highly environmentally and very economically effective.

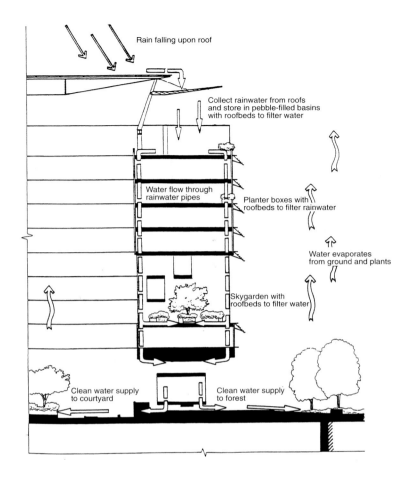

Fig. 99 Non-potable recycled water recticulation system (Source: Yeang)

Rain falling upon roof

Collect rainwater from roofs and store in pebble-filled basins with roofbeds to filter water

Water flow through rainwater pipes

Planter boxes with roofbeds to filter rainwater

Water evaporates from ground and plants

Skygarden with roofbeds to filter water

Clean water supply to courtyard

Clean water supply to forest

More than 280,000 litres of wastewater come from each U.S. household every year not from toilets but from sinks, showers, baths, dishwashers and clothes washers. This 'waste' can be safely and easily reused. Only treated water should be use for cooking, drinking, bathing and cleaning clothes and dishes. The wastewater from those uses, also called graywater, can be used to flush toilets and provide water for landscaping. Before being used for irrigation, graywater should be passed through filter systems. Even 'black water' (sewage from toilets) can be treated and used to water landscape vegetation. Many plumbing codes have not allowed the use of graywater because of concerns for sanitary conditions. Typically, separate lines and septic systems must be installed to keep gray- and black water apart. In new construction, this is not difficult, but when working with existing systems it can be costly.

Cisterns and catchment basins are ancient methods of meeting a building's water supply needs. Typically, a system of gutters from the roof collects runoff and channels it into a cistern. Ecologically conscious buildings employ these methods of collecting rainwater to reduce the need for treated water. Catchment areas are often designed to look like ponds or marshes. The rainwater collected by these methods is used for landscape maintenance. Often comparable in cost to drilling a well, rainwater collection systems can be employed within a group of buildings or an entire community. In some municipalities, rainwater may be used as a backup supply connected to the area's regular water system.

Employ gravity water storage when possible, as each foot of elevation of a storage tank provides 300,000 Kg/m² of static pressure.

Productive-mode Operational Systems

Following the assessment of the above modes in the design of our building's operational systems, the productive mode should be considered and adopted where costs permit. Productive modes are locally-generated energy sources such as photovoltaics (PV), which provide a clean, quiet and pollution-free energy source. The three main types of PV cells being produced today are single-crystal, poly-crystalline and amorphous cells. Different types of photovoltaic cells have different efficiency coefficients. These efficiencies have been calculated (Daniels, 1995) and are listed below:
- monocrystalline silicon cells
  efficiency coefficient 14 percent,
  direct voltage approx. 0.48 V,
  direct current approx. 2.9 A.

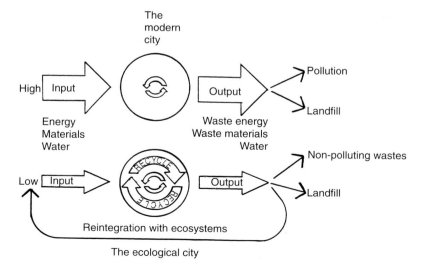

Fig. 100   Ecological cities have low inputs and low outputs (Source: Yeang)

Fig. 102 Solid-waste
recycling system

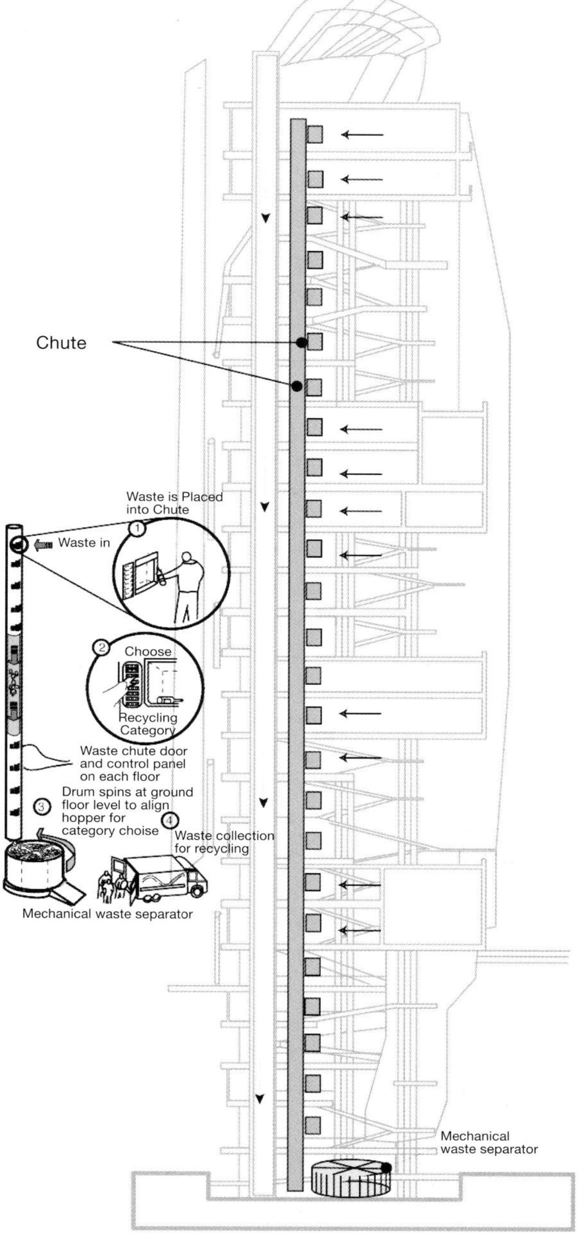

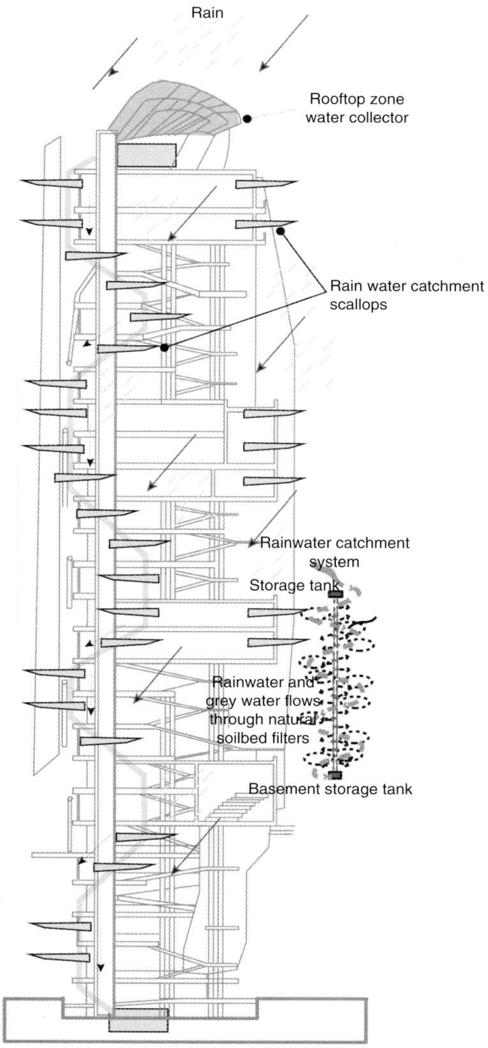

Rain

Rooftop zone
water collector

Rain water catchment
scallops

Chute

Waste is Placed
into Chute

Waste in

Choose

Recycling
Category

Waste chute door
and control panel
on each floor

Drum spins at ground
floor level to align
hopper for
category choise

Waste collection
for recycling

Mechanical waste separator

Rainwater catchment
system

Storage tank

Rainwater and
grey water flows
through natural
soilbed filters

Basement storage tank

Mechanical
waste separator

Fig. 101 Rainwater
collection and recycling
system

- polycrystalline silicon cells
  efficiency coefficient 12 percent,
  direct voltage approx. 0.46 V,
  direct current approx. 2.7 A.
- amorphous solar cells – silicon plated onto
support substance
  (opaque module)
  efficiency coefficient 5 percent,
  direct voltage approx. 63 V,
  direct current approx. 0.43 A.
- amorphous solar cells – semi-transparent
  efficiency coefficient 4 percent,
  direct voltage approx. 63 V,
  direct current approx. 0.37 A.

While single-crystal solar cells use expensive
and energy-consuming semiconductor-grade
silicon, polycrystalline and amorphous cells do
not. Polycrystalline cells use a metallurgical-
grade silicon, which is much cheaper. However,
both single-crystal and polycrystalline cells use
blocks of silicon, which must be sliced into thin
wafers to create the cells. This slicing creates
much waste in the form of dust and is a slow,
energy-intensive process.

Amorphous PV cells' manufacturing processes use far less material than others and the
product can be applied as a thin film to a
variety of materials. Amorphous PVs have traditionally had an efficiency of 5–8 percent as
opposed to crystalline panels of 10–15 percent.
However, encouraging results of 11–15 percent
efficiency have been shown in recent thin-film
panels. The development of amorphous panels
has led to architecturally integrated solar cells. The solar panel now
can be combined with part of a traditional building material,
reducing one of the costs of inputs into the construction process
and replacing it with a pollution-free, electricity producing material. The overall consequence is that PV becomes more affordable.

Being connected to the main power-grid gives a PV user the
advantage of not having a bank of costly batteries for electricity
storage. The other advantage is the reduction in the need for new
transmission lines, as distributed electricity generation would enable power to be used at its source. PV cells are usually connected

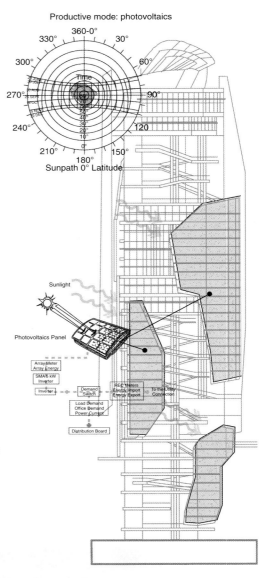

Fig. 103 **Use of ambient
energy (photovoltaics)**

in the form of modules, which are wired such that the electrical power can be extracted at a suitable voltage potential. The main component of the solar PV cell array is the power inverter. The inverter is a key component of the PV power production; it transforms the direct current produced by the solar cells to AC current at grid voltage.

In summary, the preceding chapter discusses some of the low-energy and low-ecological impact systems currently being used in buildings. It is likely that some of these will become obsolete or out-moded as new solutions and techniques emerge in this rapidly developing field.

The above discussion, however, has not evaluated these systems in terms of their embodied energy in production for comparative purposes, their life-cycle implications, their outputs of waste (such as heat and carbon dioxide) into the environment and other pertinent factors as a consequence of their usages. All of these need to

Currently, solar cells are generally expensive and of low efficiency, although current prices are falling and make their use more justifiable economically. Efficiencies are also improving. For a typical family of four in Germany using normal inefficient 220-volt AC appliances, at least 30 sq.m. of solar cells are required. It is contended here that ecological design can productively contribute to the environment, and productive mode systems such as photovoltaics are one way to make a positive contribution. Their use therefore should be encouraged.

be considered in a more rigorous approach to ecological design (as shown in the partitioned matrix in chapter 3).

In considering the operational systems in the skyscraper, the order of priority should be based on the following:
- priority given to passive systems of energy use and recycling, with such passive-mode systems being maximised, and where possible, the use of productive-mode systems.
- the adoption of those mixed-mode and active-mode systems that are low energy users and have low environmental impacts (e.g., low emissions).
- the management of the building's operational systems (whether passive, mixed-mode, productive, active or composite) as they affect the overall thermal and ecological performance and impact of the skyscraper and other intensive building types.

Referring back to the interactions matrix, any operational system adopted should be evaluated for its embodied energy and ecological impact arising from its production, its outputs arising from its operations and its recovery at the end of its useful life.

Discussion

This section serves to provide a brief overview of the ecological approach to a building's operational systems. The key principle is to start by optimising the passive-mode systems offered by the building locality's climatic conditions. This is followed by investigating the appropriate mixed-mode systems to be used, which is again dependent upon the locality's climate, on the building usage and the building's configuration and design concept. Where inevitable, full-mode systems are to be used, but adopting the energy conservation and materials recovery systems appropriate for the locality. Where affordable, productive-mode systems should be used as those generate on-site energy and localised materials recovery.

The technical aspects of the building's operational systems are not discussed in great detail as these are better covered elsewhere. The type of operational systems utilized would further depend upon the building's usage (whether residential, office, retail, etc.). Very often, the technical solutions for one type of use may conflict with another due to the different activity and time of use. Systems also depend upon the climate of the locality of the building and upon the extent to which the users are prepared to endure a less consistent level of comfort (e.g. enabling a greater use of the passive-modes or mixed-modes of operational systems).

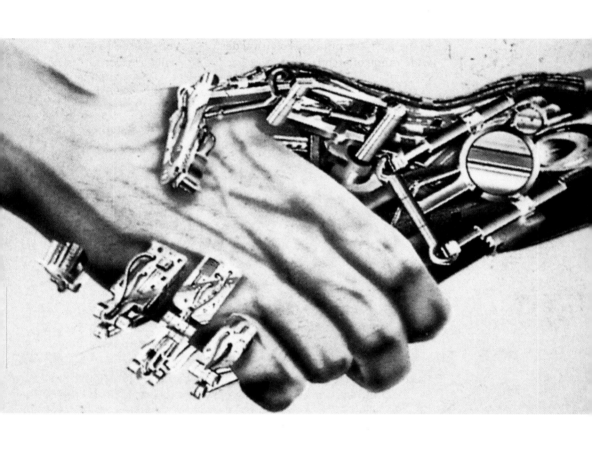

Ecological Design

In the past, the phrase 'ecological design' has been used to cover a wide range of architectural approaches. Basically, any design framework that showed any concern for environmental effects, in no matter how peripheral a fashion, has been termed ecological. But if designers are really to be sensitised to the local, regional and global environmental effects of their work, it is time to develop more rigorous criteria for ecological design. That has been the effort of this book.

Some of the basic premises of 'green' design should be obvious to all, and yet buildings continue to be produced without these premises in mind.

They include the truth that it is necessary for human beings to preserve their environment in a biologically viable state; that the current pace of environmental destruction wrought by human actions cannot continue without disastrous consequences and must be halted; and that the negative ecological impacts of our actions have to be minimised as much as possible. Simply stated, ecological design requires that the designer take concerted and coherent action throughout the design process to mitigate the anticipated negative effects of his or her intervention in the natural environment (the creation of the building). From start to finish – and by finish is meant the end of the building's useful life – green design takes into account effects on the earth's ecosystems and resources, and continuously tries to eliminate or mitigate adverse impacts while also trying to increase the design's positive and reparative effects on the environment. For example, the designer may simultaneously eschew the use of rare hardwoods, cut the building's energy consumption by a significant percentage and reintroduce native plants to a previously 'zero-cultural' urban site. Figures 101 and 102 illustrate the conceptual framework the designer uses to reduce negative ecological effects.

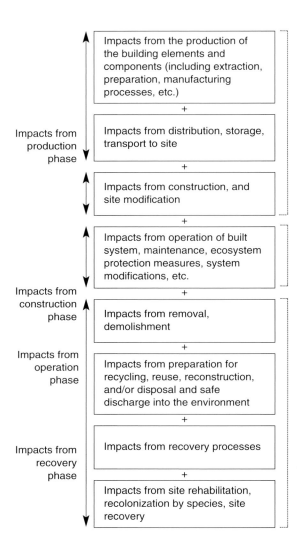

Fig. 104   Impacts in the
life cycle of a designed system (Source: Yeang, 1995)

But before the designer can do any of these things, he or she has
to be able to identify those features of the built environment that
cause harm to the natural one. A thoroughly ecological approach,
therefore, makes use of the framework specified by the partitioned
matrix discussed in chapter 3 and the full set of interactions which
specify the totality of interrelations between the building and the
biosphere (labeled L11, L12, L22 and L21 in the matrix).

Environmental problems related to built structures are those
changes in the ecosystem resulting from the production, operation
and final disposal of the building; therefore, they include the entire
life cycle of the built environment, for which the designer becomes

Fig. 105   Ecological
design model (Source:
Yeang, 1995)

responsible. Among these negative impacts are the depletion of terrestrial resources,
reduction of biological diversity and ejection of pollutants
into the ecosystem. Ecological
design should minimise such
changes, whether they occur
through depletion, alteration
or addition to the earth's ecosystems and the project site. To
do so, an interdisciplinary
approach is required, since the
effects of the built environment overlap many areas.
Environmental protection,
waste disposal, materials science and of course ecology are
all involved. One of the reasons why earlier 'ecological design' has
been so haphazard and incomplete is that most designers lack the
knowledge of ecology, biology and other fields that impinge on environmental issues. The theory presented in this book provides a basis
for an integrated design framework so that the complex set of interdependencies will be factored into the design; otherwise, 'ecological
design' will continue to be piecemeal or linear. Therefore, the interactions framework is the centerpiece of a holistic design theory.

   We should also be aware that with so many emerging fields of
knowledge and study (e.g. resource conservation, pollution control
and low-energy engineering), the criteria for more ecologically
informed design will become increasingly complex. This will put
further pressure on the practising designer or architect who is
already faced with more information available to solve a problem
than he or she can possibly assimilate. In order to proceed, design
must be selective and respond to the most significant issues within
the context of the design problem.

The overall ecological design
approach should be based on the
model:

$$(LP) = \frac{\begin{array}{c|c} L11 & L12 \\ \hline L21 & L22 \end{array}}{}$$

where

L11 Refers to the processes and
     activities that take place within
     the system

L22 Refers to the processes and
     activities that take place in the
     environment of the system

L12 Refers to the exchanges of the
     system with its environment

L21 Refers to the exchanges of the
     environment with the system

Education and Research

Architecture is a social art, in which a large number of disciplines
already have a hand. Ecology is a natural science, and includes the
question of human survival within the natural environment – an
environment that humans are constantly using and changing,
often in ways that go against the sustainability of the species.

Building is one of the primary ways in which people change the natural environment, not only by displacing it at the site, but by using resources from far and near in constructing the building. It logically follows that designers and architects, who are responsible for the creation of built structures, should increase their knowledge of the interactions between architecture and the environment if we are to reduce the negative impacts of human constructions on the natural systems with which we share the biosphere. The teaching and practice of architecture will therefore have to be modified to include an ecological analysis of the building, use and disposal of human built structures.

The need for ecological design is no longer seriously in dispute, but a comprehensive 'green' approach has yet to be embodied in architectural practice. Current theory tends to focus on two camps, which promote the spatial (e.g., Martin, 1967) and climatic models (Hillier, 1977), respectively. A truly ecological approach would include both. Any built structure obviously causes a spatial displacement in its environment simply by existing, but it also alters the climate through its operation.

Ecological design goes further, to include the interactions of architecture and environment, which go beyond spatial displacement and effects at the project site. Thus, the types and quantities of energy and resources that go into the building's structure, their source and the environmental effects at the point of extraction all need to be considered. In terms of functioning, the internal processes of the built environment and its outputs also have to be examined, and, in turn, the responses or 'feedback' of the ecosystem as a result of its efforts to absorb the outputs. The four components in the interactions matrix correspond to the sets of demands that any built environment makes on the earth's ecosystems. Each is related

Traditional architectural education, obviously, will have to be modified. Ecology and environmental biology will have to be included in architectural curricula. In addition, related disciplines such as resource conservation, recycling, energy and materials management and pollution control are also germane to ecological design and must also be taught.

to other disciplines, such as those concerned with pollution, environmental protection and conservation. Therefore, the framework provides us with a fundamental basis from which research on ecological design and all problems of environmental impairment could be approached.

The ecologist is more concerned with the systemic aspects of architecture than with its aesthetic or social dimensions (even though these aspects may indirectly have ecological implications, in the sense that they affect the behaviour of people in the ecosystem). The consumption of energy and materials and the flow from sources to environmental sinks is of more interest to the ecologist, for whom a building is a machine that consumes resources in the transient form of the built environment located at a particular project site, but which has a connection to large-scale effects in the biosphere. Ecologically speaking, the building is also a potential waste product, which has to be recycled just as a soda can or a plastic bottle. Designers may resist viewing their artworks in this manner, and cling to the traditional view of the building as an outgrowth of the architect's aesthetic and its functional use. However, a building is both of these things, a structure with a particular social and economic function that embodies the aesthetic theories of an individual or group of individuals, and also requires certain resources to be realised and creates certain environmental effects as a result of its creation and operation.

The traditional conception of the architect's professional responsibilities will also be expanded. For instance, if the designer is aware of the implications of using certain forms of energy and materials that may be toxic, he or she is responsible for the resulting ecological effects. The designer is, from a green perspective, responsible for the choice of materials and systems *and* for the way the structure and its components are used, recycled or discarded at the end of the building's life. Again, architects may resist the idea that their buildings will one day be torn down and replaced, preferring to think of them in a timeless aesthetic way; but such a view is not in keeping with an awareness of the need for sustainable building practices and the conservation of the environment and natural resources. Obviously, the amount of resources consumed in the creation of a large or tall urban building is huge; whether this material eventually becomes waste will largely be determined by design choices before the building is even raised, and is therefore the designer's responsibility. More research needs to be done on the ecological impact of energy and materials use. But if the responsibility that is here being loaded on the architect's shoulders seems enormous, it has to be remembered that it is also an enormous opportunity to direct and determine how human beings interact with the natural environment and whether they behave in a way that makes their very existence sustainable.

The building will also need to be monitored throughout its life cycle for ecological effects, and a system will have to be put in place for doing so. Thus, the architect's job is not over once the building has risen over the project site, but continues from 'source to sink'. Much literature has accumulated on design methods (e.g., Jones, C., 1967) and a variety of approaches have been suggested. In approaching a design problem, the architect has to strike a balance that assigns a relative importance to the various elements, and may use this approach in one case, that approach in another. This case-by-case method, using various approaches as the environmental demands dictate, is an effective method of designing responsibly and ecologically. Different approaches have different strengths and weaknesses; as technology and systems advance and are further developed, the costs and benefits will change over time. Therefore, it is impossible to pre-select a set of standard solutions for design problems; all we can do is give the architect a philosophy and a method of working that is itself renewable and sustainable and responds to each situation with the best knowledge currently available in order to create an effective design synthesis.

As we are still in the early days of ecological design and its technology, we can only make predictions about the design options a designer will have for consideration. The interactions specified in the partitioned matrix will always be relevant, but the technical solutions to the ecological problems will evolve. Initial design decisions are vitally important; as has been mentioned, they will determine the degree of environmental dislocation and destruction and the extent and feasibility of preventive or ameliorative actions.

The fact that human beings change the environment by their behavior does not have to be inherently destructive. The quantification of the impacts defined by the partitioned matrix gives no more than a statement of the extent of impacts on ecosystems and the corresponding resources needed to mitigate them. The designer's task is to integrate the skyscraper or our intensive building type with its ambient ecosystem to minimise negative effects and by means of design interventions to achieve a 'steady-state' relationship with the environment (Girardet, 1992; World Resources Institute, 1994). Further research needs to be done to develop quantitative data to support ecological design, covering the areas of interaction specified in the partitioned matrix (see Fig. 105) and extending over the building's entire useful life. Some of the elements that need quantitative development are the quantities of energy and materials used in the building; global availability of these resources and their rates of depletion; ecological consequences of each input; permissible levels of outputs by the building and the routes taken by these discharges through the ecosystem; the energy and materials cost of outputs management; the appropriateness and efficiency of the building's operational systems; the extent of internalization of system processes; the ecological consequences of the operational systems; biodiversity of the building's ambient ecosystem, and resilience; the global impacts on natural systems of the building; impacts on other manmade systems; and the global impact on renewable and non-renewable resources.

Ecological design does not mean that the entire biosphere should be isolated from human intervention and turned into a nature preserve. Ecosystems change whether humans intervene in them or not; the goal of green design is manage the interaction of people and environments in the least harmful way possible, taking into account the limitations of ecosystems and biosphere resources and managing their use in a sustainable manner. In principle, the green skyscraper and the intensive building type could actually have beneficial ecological impacts. Critical design choices will determine whether positive effects are achieved, and how. An organized and coherent use of ecological principles in design is still under development; unfortunately, in many cases the information, theory, and practical applications to achieve the desired effects are not yet available. Ecosystem modification by built systems has so far been largely ignored or misunderstood, and unforeseen effects are not uncommon. Ecological design will require transforming existing buildings and creating new ones to function, both individually and collectively, in tune with natural systems. The question then is whether we can understand the complexity of the natural world well enough to emulate it at all successfully; for example, our current methods of understanding nature still rely on classical, reductive science. Dissertation, for instance, will never capture the synergies and intricate interrelations that distinguish the systems in the natural world (Cole, 1995). It is essential that more systematic endeavours be made by researchers to achieve a deeper understanding of environmental repercussions, to anticipate them as a regular part of design and planning and to collect the data in a form appropriate to design analysis.

The gaps in our knowledge and practice should not dissuade the designer from incorporating ecological considerations into design. When data cannot be exactly quantified, indices can be used (such as the water-quality index). Better empirical data should follow. The provision of monitoring systems to facilitate environmental protection systems is essential. Here again, more data are needed, since many potentially dangerous pollutants have not even been recognised or evaluated yet. The four components of the interactions framework also need to be further studied and reliable data developed on inputs and outputs of the built environment, the interactions of the systems of the ecosystem and the built environment and other factors.

Of course, the larger social pattern of use and the pressure or desire for sustainable resource consumption and conservation are elements that go beyond the scope of the designer's power. But an ecologically minded designer can at least 'buy time' until society has developed more ecological consumption habits and more responsible value systems and styles of living – just as the designer has to wait for the development of new environmentally responsive technologies but can use the best that are available in the meantime (which can be made available through rapid prototyping) and make allowances for future developments. The pursuit of green design is itself an impetus for the changing of people's ideas about the relationship of their built environment to the natural world, and is also a spur toward the development of new, 'green' technologies. There is no reason why architects cannot lead these developments rather than waiting for them to happen, and every reason for them to start making all buildings as green as possible.

To conclude, if we are to ensure the built environment's long-term survival in the biosphere (and, of course, that of its inhabitants), then its systems must have certain obligatory relationships with the ecosystem and its processes. Rather than being designed to be totally isolated from the ecosystems (which in any case is impossible), the built environment and the intensive large urban building in particular should be integrated and have compatible symbiotic relationships with ecosystems. In occupying any location, a designed system has at the same time assumed an ecological role (whether contributory or passive) in relation to that ecosystem's composition and function. In the ecological approach, this role is accepted and made part of the designer's responsibility. The architect must ensure that his or her creation responds as an integral component of the ecosystem; this means it is not seen as an isolated structure severed from the project site's ecosystem, nor made completely dependent (parasitic) on the surrounding ecosystems. While difficult to generalise for all sites, impacts need also to be prioritised (Cole, R., et al. (eds.), 1995).

It is clear that the architect and designer play a central role in ensuring a sustainable future. Papanek (1985) asserts that, 'In this age of mass production when everything must be planned and designed, design has become the most powerful tool with which man shapes his tools and environments (and by extension, society and himself). This demands high social and moral responsibility from the designer'. Hayes (in Peters, T., 1997) makes a similar reference to the importance of design in business competitive strategy: 'Fifteen years ago, companies competed on price. Today it is quality. Tomorrow it is design...'

What is described in this book is a set of ecological ideals or intentions, the full implementation of which may incur additional costs (over and above the conventional building costs) or societal changes (e.g. standard-of-living or comfort) or the revision of current technological and design methods. While many of the ecological objectives remain currently technologically or scientifically unfulfilled, the framework described here nevertheless provides a point of departure from which, hopefully, these objectives will eventually be achieved in their entirety.

A final word about aesthetics, economics and performance: at the beginning of this work, I asserted that it is the intensive and the large-building types that particularly need the attention of the world's ecological designers to make them ecologically responsive as possible, for the sake of our sustainable future. However, we might conclude here by declaring that in addition to meeting the systemic aspects of ecological design, the ecologically responsive or 'green' skyscraper or large building type must also be aesthetically pleasing, economically competitive and excel in performance. If it does not meet these criteria, it is likely that it will not be accepted by the public. The economics of ecological design (or ecological economics) need to be rationalised if business is to accept the benefits of green design.

According to McDonough, W. (in Zeiher, 1996, p. 49), '... the solar efforts of the 1970s failed because it wasn't beautiful, ultimately it didn't work and often didn't even cost out'. This concern is similarly voiced by Wells (1984, p. 47), that '... if ever we needed great designers, it is now. The environmental architecture of America is almost without exception depressingly ugly...' Low-energy design and ecological design are applicable regardless of architectural style. Since the best opportunity for improving a building's environmental performance occurs early in the design process, it is clear then that we must at the onset make our skyscrapers and other large buildings not only ecologically responsive but aesthetically pleasing as well if green design is to be a durable proposition.

References

Albuquerque-Beralillio County Planning Dept. *Comprehensive Plan, Metropolitan Environmental Framework*. Albuquerque, 1972, p. 52.

Amato, A., and A. Eaton. "Life Cycle Assessment." Paper presented at the Steel Construction Institute's "Sustainable Steel" conference. Orlando, Fla., 18–21 March, 1998.

Ambasz, E. "The Formation of a Design Discourse." *Perspecta: The Yale Architectural Journal*, vol. 12, 1969, pp. 57–70.

Arvil, R. *Man and Environment*. London: Penguin Books, 1970.

ASCE. *Proceedings of the ASCE Urban Transportation Division Environmental Impact Specialty Conference, May 21–23, 1973, Chicago, Illinois*. New York: American Society of Civil Engineers, 1973.

Ashby, F. *Royal Commission on Environmental Pollution, First Report*. London: H.M.S.O., 1971.

Ashby, F. *Royal Commission on Environmental Pollution, Second Report: Three Issues in Industrial Pollution*. London: H.M.S.O., 1972.

Ashby, F. *Royal Commission on Environmental Pollution, Third Report: Pollution in some British Estuaries and Coastal Waters*. London: H.M.S.O., 1973.

Ashby, R. *Introduction to Gybernetics*. London: Chapman & Hall, 1956.

Atkins, J. Thomas, et al. *Huntington Environmental Planning Program*, M.A. thesis, Dept. of Landscape Architecture and Regional Planning, University of Pennsylvania, 1972.

Audubon Society Hq. in L.C. Zeiher, *The Ecology of Architecture: A Complete Guide to Creating the Environmentally Conscious Building*. New York: Whitney Library of Design, 1996.

Aynsley R. "Natural Ventilation in Passive Design" in *RAIA Environment Design Guide*, May 1998, TEC 2, RAIA Publishing, Australia.

Baggs, S.A. "Underground Architecture." *Architecture Australia*, vol. LXVI, no. 6, December 1977, pp. 62–9.

Baker, N.V. *Energy and Environment in Non-Domestic Buildings: A Technical Guide*. Cambridge: Cambridge University Press with Cambridge Architectural Research Ltd. and The Martin Centre for Architectural and Urban Studies, 1992.

Ballard, D.W. "An American View of Problems of Materials Conservation." *Conservation of Materials*, report, Harwell, England Conference, 1974, pp.15–37.

Barrows, H.H. "Geography as Human Ecology." *Annals of the Association of American Geographers*, 13, 1973, pp.1–14.

Bateson, G. *Steps to an Ecology of the Mind*. London: Paladin, 1973.

Beckman, M., and J. Weidt. *Archiecoframe: or, A Partial Analysis of the Ecological Impact of Building Materials and Residential Scale Structural Systems*. Minneapolis: Dept. of Architecture, University of Minnesota, 1973.

Behling, S., and S. Behling. *Sol Power: The Evolution of Solar Architecture*. Munich: Prestel Publishing, 1996.

Bender, Tom. *Environmental Design Primer*. New York: Schocken Books, 1973.

Berdurski, B.L. "Ecology and Economics–Partners for Productivity." *The Annals of the American Academy of Political and Social Sciences*, vol. 410, 1973, pp. 75–94.

Berry, R. "Recycling, Thermodynamics and Environmental Thrift." *Bulletin of Atomic Science*, vol. XXVIII, May 1972, pp. 8–15.

Bertalanffy, L. von. *General Systems Theory: Foundations, Development, Applications*. New York: George Braziller, 1968.

Billings, W.D. *Plants and the Ecosystem*. Belmont, Calif.: Wadsworth, 1964.

Bolitho, V. "Environmental Inter-Relationships in Water Pollution Control." *Water Pollution Control*, 1973.

Bookchin, M. "Environmentalists versus Ecologists." *Undercurrents* 4, Spring 1973, London, Geoffrey Boyle.

Bormann, F.H.G., and G.E. Liken. "Nutrient Cycling." *Science*, vol. 155, 1967, pp. 424–429.

Boughey, A.S. *Fundamental Ecology*. London: Interext Books, 1971.

Boulding, K. *Economics of the Spaceship Earth*. New York: Harper & Row, 1969.

Bowen, H. "Design for Survival." *Architectural Design*, 1972, p. 418.

Bower, B.T., et al. *Waste Management, Generation and Disposal of Solid Liquid and Gaseous Wastes in the New York Region, A Report of the 2nd Regional Plan*. New York: Regional Planning Association Publications, 1968.

Bower, B.T., and W.O. Spofford. "Environmental Quality Management." *Natural Resources Journal*, vol.10, no. 4, October 1970, pp. 655–667.

Bower, B.T. "Interpretation: Residuals and Environmental Management." *AIP Journal*, July 1971, pp. 218–220.

BRECSU (Building Research Energy Conservation Support Unit). *Energy Consumption Guide No. 19*, Hong Kong, 1995.

Brooks, H., and R. Bowers. "The Assessment of Technology." *Scientific American*, vol. 222, no. 2, 1970, pp.13–22.

Brown, G., and P. Stetlon. "The Energy Cost of a House." *Rational Technology Unit at the Architectural Association, 1973–74*. London: AA Publishers, 1974.

Brown, H. "Human Materials Production as a Process in the Biosphere." *Scientific American*, vol. 222, no. 3, Sept. 1970, pp. 194–209.

Brown, Lester. *Building a Sustainable Society*. New York: Norton, 1981.

Brown, Lester, Christopher Flavin, and Sandra Postel. *Saving the Planet*. New York: W.W. Norton, 1991.

Brubaker, S. *To Live on Earth: Man and his Environment in Perspective*. London: Johns Hopkins Press for Resources for the Future, 1972.

Brundtland, G.H., in World Commission on the Environment and Development (WCED). *Energy 2000: A Global Strategy for Sustainable Development*. London: Zed Books, 1987.

Burall, P. *Green Design*. London: The Design Council, 1991.

Cain, G. "The Ecological House." *Street Farmer*, 1. London: AA Publishers, 1971.

Cain, S.A. "The Importance of Ecological Studies as a Basis for Land-Use Planning." *Biological Conservation*, vol. 1. Great Britain: Elsevier Publishing Co., 1970.

Carolin, P., and T. Dannat. *Architecture, Education and Research: The Work of Leslie Martin: Papers and Selected Articles*. London: Academy Editions, 1996.

Carr, Marilyn, ed. *The AT Reader: Theory and Practice in Appropriate Technology*. London: Intermediate Technology Publications, 1985.

Celedrovsky, G. "Systems in Balance." Office of Architecture and Planning, March 1970.

Chanlett, Emil T. *Environmental Protection*. New York: McGraw-Hill, 1973.

Chapman, P. *The Energy Cost of Producing Copper and Aluminum from Primary Sources*. Open University Report ERG 001, Rev. Doc., 1973.

Chapman, P. "The Energy Cost of Producing Copper and Aluminium from Primary Sources." *Metals and Materials*, Feb. 1974, pp.107–111.

Chappell, C.L. "Disposal Technology for Hazardous Wastes." *Surveyor*, Nov. 2, 1973, pp. 501–502.

Chermayeff, S., and A. Tzonis. *Shape of Community: Realisation of Human Potential*. Hammondsworth: Penguin Books, 1971.

Clarke, R. "Soft Technology: Blueprint for a Research Community." *Undercurrents* 2, May 1972.

Clements, F.E. *Plant Succession: An Analysis of the Development of Vegetation*. New York: Carnegie Institute, pp. 242, 515.

Clinic for Occupational Medicine, Orebro, Sweden. "The Healthy Environment." *Southeast Asia Building*, November 1994, p. 54.

Cole, Ray, et al. *Linking and Prioritising Environmental Criteria, Proceedings*. Task Group 8, International Research Workshop, Toronto, Canada, November 15–16, 1995. School of Architecture, University of B.C., 1995.

Colinvaux, Paul R. "The Ecosystem as a Practical Model" in *Introduction to Ecology*. New York: John Wiley & Sons, 1973, pp. 229–245.

Common, M. "Economics and the Environmental Problem." *Discussion Papers in Conservation*, no. 5. London: University College, 1973.

Commoner, Barry. *The Closing Circle, Confronting the Environmental Crisis*. London: Jonathan Cape, 1972.

Cook E. "The Flow of Energy in an Industrial Society." *Scientific American*, vol. 225, no. 3, Sept. 1971, pp. 134–147.

Cooke, G. D., R. V., and E.P. Odum. "The Case for the Multispecies Ecological System with Special Reference to Succession and Stability." *Biogenerative Systems*, NADA Spec. 165, 1968, pp. 129–130.

Costin, A.B. "Replaceable and Irreplaceable Resources and Land Use." *Journal of the American Institute of Agricultural Science*, March 1959, pp.3–9.

Cranbrook, Earl of, and D.S. Edwards. *A Tropical Rainforest: Biodiversity in Borneo at Belalong,*

Brunei. London: The Royal Geographical Society, 1994.

Crosbie, M.J. *Green Architecture: A Guide to Sustainable Design.* Rockport, Mass: Rockport Publishers, 1994.

Crosby, T. *The Environmental Game.* London: Penguin Books, 1973.

Dales, J.H. *Pollution, Property and Prices: An Essay in Policy Making and Economics.* Toronto: University of Toronto Press, 1968.

Daly, Herman. *Steady-State Economics.* Washington: Island Press, 1991.

Daniels, Klaus. "Klima und Gebäudeform" in *Technologie des Ökologischen Bauens.* Berlin: Birkhäuser, 1995, pp. 18–33.

Darling, Fraser F. *Wilderness and Plenty.* Reith Lectures, BBC, 1969.

Darling, Fraser F., and R.F. Dasmann. "The Ecosystem View of Human Society." *Realities,* June 1972.

Darling, Fraser F., and J.P. Milton, eds. *Future Environments of North America.* Garden City, N.Y.: The Natural History Press, 1966.

Dasmann, R.F. *Environmental Conservation.* New York: Wiley, 1968.

Dasmann, R.F. "Towards a System for Classifying Natural Regions of the World and their Representation by National Parks and Reserves." *Biological Conservation,* vol. 4, July 1972, pp. 247–255

Dasmann, R.F., J.P. Milton, and P.H. Freeman. *Ecological Principles for Economic Development.* London: John Wiley and Sons, 1973.

Davoll, J. *Statement by the Conservation Society and the Minister of Environment on the U.N. Stockholm 1972 Conference,* London, 1972.

Dee, N., I.L. Whitman, J.T. McGinnis, and D.C. Fahringer. *Final Report on the Design of an Environmental Evaluation System to the Bureau of Reclamation, U.S. Dept of the Interior.* 14-06-D-7005, June 30,1971, Columbus, Ohio, Battelle, Mem. Inst. Publ., 1971.

Demkin, J.A., ed. *Environmental Resource Guide* (ERG), The American Aluminum Institute of Architects. New York: John Wiley & Sons, Inc., 1996.

Desmecht, J.,A. Dupagne, J. Hauglustaine, and J. Teller. "Low-Energy Social Housing, Marchin (Belgium): A Case-Study Revisited." *Environmentally Friendly Cities, Proceedings of PLEA '98.* (Lisbon, Portugal, June 1998). London: James & James Science Publishing Ltd., 1998, pp. 237–240.

Detwyler, T.R. *Man's Impact on the Environment.* New York: McGraw-Hill, 1971.

Detwyler, T.R., and M.G. Marcus, eds. *Urbanization and Environment: the Physical Geography of the City.* Belmont, Calif.: Duxbury Press, 1972.

Diamant, R.M.E. "The Prevention of Pollution." *H & VE,* 1971.

Dickson, David. *Alternative Technology and the Politics of Technical Change.* London: Fontana, 1974.

Dubin, F.S., et al. *Energy Conservation Design Guidelines for New Office Buildings.* Washington Report for General Services Adminstration/Public Building Service, 1974.

Duffey, E., and A.S. Watt. *The Scientific Management of Animal and Plant Communities for Conservation.* The 11th Symposium of the British Ecological Society, University of E. Anglia, Norwich, July7–9. Oxford: Blackwell, 1970.

Dunn, J.B., and J.A. Hington. *The Re-Vegetation of Despoiled Land.* Public Works and Municipal Services Congress, no. 14. London: Oyez Press, 1970.

Dunn, P.D. *Appropriate Technology: Technology with a Human Face.* London: Macmillan, 1978.

Durrell, Lee. *Gaia State of the Ark Atlas.* New York: Doubleday, 1986.

Edwards, Brian. *Towards Sustainable Architecture, European Directives and Building Design.* Oxford: Butterworth Architecture, 1996.

Edwards, Brian. *Green Buildings Pay.* London: E&FN Spon, 1998.

Egler, F.E. "Vegetation as an Object of Study." *Philosophy of Science,* vol. 9, no. 3, 1972, pp. 245–260.

Ehrenfield, O.W. *Biological Conservation.* New York: Holt, Reinhart and Winston Inc., 1970.

Ehrlich, Paul R., and H. Anne. *Population Resources Environment: Issues in Human Ecology.* San Francisco: W.H. Freeman & Co, 1970.

Emery, F.E., ed. *Systems Thinking: Selected Reading,* Part 4. Harmondsworth: Penguin Books, 1981, pp. 12, 241–260.

Emery, F.E., and E.C. Trist. "The Casual Texture of Organizational Environments." *Human Relations* 18, 1965, pp. 21–32.

Fathy, Hassan. *Natural Energy and Vernacular Architecture.* Chicago: published for United Nations University by University of Chicago Press, 1986.

Fazal, A. "The Future of Cities: Sustainable Settlements in Asia Pacific" in A. Awang et al., *Towards a Sustainable Urban Environment in Southeast Asia: Urban Habitat and High-Rise Monographs SEACEUM2.* Kuala Lumpur: Institute Sultan Iskandar of Urban Habitat and High-Rise (ISI), 1995, p. 3.

Flajser, S.H., and Porter, A.C. "Towards a Science for Technology Assessment." *The Trend in Engineering*, vol. 25, no. 2, April 1974.

Flavin, Christopher. *Energy and Architecture: The Solar and Conservation Potential*. Washington, D.C.: Worldwatch Institute, 1986.

Flavin, C., and N. Lenssen. *Power Surge: Guide to the Coming Energy Revolution*. New York: W.W. Norton & Company, 1994.

Flawn, P.T. *Environmental Geology: Conservation, Land-Use Planning, and Resource Management*. New York: Harper and Row, 1970.

Fordham, Max. "Thinking Big." *Architectural Review*, August 1997, pp. 87–89.

Fox, A., and R. Murrell. *Green Design*. London: Architecture Design and Technology Press, 1989.

Fuller, R.B., et al. *World Design Science Decade 1965–1975*. Carbondale, Ill.: Southern Illinois University Press, 1963.

Gabel, M. *Energy, Earth & Everyone*. New York: Simon & Schuster, 1975.

Gerardin, Lucien. *Bionics*, London: Weidenfeld & Nicolson, 1968.

Ghazali, Z.M., and M.A. Kassim. "How to Reduce Wastes and Save Materials" in A. Curarg et al., eds., *Environmental and Urban Management in Southeast Asia, Urban Habitat and High-Rise Monographs*. Kuala Lumpur: Institute Sultan Iskandar of Urban Habitat and High-Rise (ISI), 1994, pp. 183–195.

Girardet, H. *Earth Rise: How Can We Heal Our Injured Planet?* London: Paladin, 1992.

Givoni, B. *Passive and Low-Energy Cooling of Buildings*. New York: Van Nostrand Reinhold, 1994.

Goldbeck, N., and D. Goldbeck. *Choose to Reuse*. Woodstock, N.Y.: Ceres Press, 1997.

Goldsmith, E. "Limits of Growth in Natural Systems." *General Systems* 16, 1971.

Goldsmith, E. *Can Britain Survive?* London: Sphere Books Ltd., 1972.

Gordon, A. "The President Introduces his Long Life/Loose Fit/Low Energy Study." *RIBAJ*, Sept. 1972, pp. 374–375

Grant, Donald P., and W.S. Ward. "A Computer-Aided Space Allocation Technique." *Proceedings from the Kentucky Workshop on Computer Applications to Environmental Design*. Lexington: University of Kentucky College of Architecture, 1970.

Graves, William, ed. "Water: The Power, Promise and Turmoil of North America's Fresh Water." *National Geographic Special Edition*, November 1998.

GSA. *Energy Conservation Design Guidelines for Office Buildings*. Washington, D.C.: Dublin-Mindell-Bloome Associates in cooperation with AIA Research Corp., 1974.

Guzowski, M., R. Horst, and S. Sorensen. *Daylight Impact Assessment, Phase One: Literature Speech for Northern States Power Company*. Report published by The Designer Group with the Weidt Group, The Regional Daylighting Center, 1994, pp. 2–7.

Hamilton, D. *Technology, Man and the Environment*. New York: Scribner, 1973, p. 211.

Hannon, B. *System Energy and Recycling: A Study of the Beverage Industry* (CAC 23). Urbana: Illinois University, Center for Advanced Computation, 1973.

Harada, Shizuo. *Private Communication*. Japan Hypertower Group, 1996.

Harper, Peter. "What's *Left* of Alternative Technology?" *Undercurrents* 6, March/April 1974, London, pp. 35–39.

Harte, J., and R.H. Socolow. *Patient Earth*. New York: Holt, Rinehart and Winston, Inc., 1971.

Hasler, A.D., et al., eds. *Man in the Living Environment*. Report of the Workshop of Global Ecological Problems. Madison: The Institute of Ecology, University of Wisconsin Press, 1972.

Hayes, R., in T. Peters, *The Circle of Innovation: You Can't Shrink Your Way to Greatness*. UK: Hodder & Stoughton, 1997, p. 429.

Hayhow, D. *Materials in Building Circles*. Dissertation presented for the Second Diploma Examination, Cambridge, Mass.: University School of Architecture, 1974.

HBI (Healthy Building International Inc.). *Building Investigation Experiences*. HBI Inc.,1994.

Hertel, H. *Structure, Form: Movement*. New York: Reinhold, 1966.

Herzog, T., ed. *Solar Energy in Architecture and Urban Planning*. Munich: Prestel, 1996.

Hillier, B. "Architectural Research: A State of Mind." *RIBAJ*, May 1977, p. 202.

Hi-Rise Recycling Systems, Miami, Florida, pamphlet.

Hirst, E. *Energy Implications of Several Environmental Quality Strategies* (ORNL-NSF-EP-53). Oak Ridge, Tenn.: Oak Ridge National Lab., 1973.

Holdgate, M.W. *Action Against Pollution*. Washington, D.C.: Central Unit on Environmental Pollution, D.O.E, 1972.

Holdren, J.P., and P.R. Ehrlich. "Human Population and the Global Environment." *American Scientist*, vol. 62, May-June 1974, pp. 282–292.

Holling, C.S., and O. Gordon. "Towards an Urban Ecology." *Bulletin of the Ecological Society of America*, June 1971.

Holling, C.S., and M.A. Goldberg. "Ecology and Planning." *AIPJ*, vol. 37, no. 4, July 1971, pp. 221–230.

Holling, C.S., and M.A. Goldberg. "The Nature and Behaviour of Ecological Systems." *An Anthology of Selected Readings for the National Conference on Managing the Environment*. International City Management Association, Washington, D.C., May 1973, pp.1–21.

Hough, M. *City Form and Natural Process: Towards a New Urban Vernacular*. London: Croomhelm, 1984.

Hough, M. *City Form and National Process*. London: Routledge, 1995.

Howard, N.J., and P. Roberts. "Environmental Comparisons." *The Architects Journal*, vol. 21, September 1995.

Howard, N.J., and H. Sutcliffe. "Precious Joules." *Building*, vol. 18, March 1997, pp. 48–50.

Hughes, M.K. "The Urban Ecosystem." *Biologist*, vol. 21, no. 3, Aug. 1974, pp. 117–127.

Hutchinson, Sir Joseph. "Land Restoration in Britain–by Nature and Man." *Environmental Conservation*, vol. 1, no. 1, Spring 1974, pp. 37–41.

Institute of Ecology. *Man in the Living Environment*. Madison: University of Wisconsin Press, 1972.

Isard, Walter, et al. "On the Linkage of Socio-Economic and Ecologic Systems." *The Regional Sciences Association Papers*, vol. 21, 1969, pp. 79–99.

Isola, A.V.D. "New Concepts of Construction Costs." *Actual Specifying Engineer*, June 1973, p. 116.

Istock, C.A. "Modern Environmental Deterioration as a Natural Process." *International Journal of Environmental Studies*, vol.1, no. 2, 1971, pp. 151–155.

Istock, C.A. "Some Ecological Criteria for the Preparation and Review of Environmental Impact Statements." *Proceedings of the ASCE Urban Transportation Division*, Environmental Impact Statements, Specialty Conference, May 21–23, 1973, Chicago, Illinois.

Istock, C. *Personal Communication*,1974.

Jeger, L. *Taken for Granted: Report of the Working Party on Sewage Disposal*. London: HMSO, 1970.

Jones, Martin V. *A Technology Assessment Methodology: Some Basic Propositions*. Washington, D.C.: Office of Science and Technology, Executive Office of the President and the Mitre Corporation, MTR - 6009, vol. 1, NTIS PB 202778-01, 1971.

Kahn, H. *The Next 200 Years*. London: Abacus, 1978.

Kaiser, E.J., et al. *Promoting Environmental Quality through Urban Planning and Controls*. Washington, D.C.: U.S. Environmental Protection Agency, Socio-Economic Environmental Studies Division, EPA-600/5/-73-015, 1974.

Karyono, T.H. "Architectural Science, Informatics and Design" in J.W.T. Kan, ed., *Proceedings of the 30th Conference of the Australia and New Zealand Architectural Science Association (ANZAScA)*. Dept. of Architecture, Chung Chi College, Chinese University of Hong Kong, July 17–19, 1996, pp. 207–211.

Kasabov, G. "Buildings: The Key to Energy Conservation." *RIBAJ*, Oct. 1979, pp. 440–441.

Klaff, Jerome L. "National Materials Policy Necessary to Conserve U.S. Resources." *Environmental Science and Technology*, vol. 7, no. 10, October 1973, pp. 913–916.

Kneese, A.V., S.E. Rolfe, and J.W. Harned, eds. *Managing the Environment, International Economic Cooperation for Pollution Control*. New York: Praeger Publishing, 1971.

Knowles, R.L. *Energy and Form, An Ecological Approach to Urban Growth*. Cambridge, Mass.: The MIT Press, 1974.

Koenig, H., W. Cooper, and J.M. Fahrey. "The Industrial Ecosytem." *Our Technological Environment: Challenge and Opportunity*. Washington, D.C.: American Society for Engineering Education, 1971.

Kormondy, E.J. *Concepts of Ecology*. Englewood Cliffs, N.J.: Prentice Hall, 1969.

Krebs, C.J. *Ecology: the Experimental Analysis of Distribution and Abundance*. New York: Harper and Row, 1972.

Kurn, D.M., S. E. Bretz, B. Huang, and H. Akbari. "The Potential for Reducing Urban Air Temperature and Energy Consumption Through Vegetation Cooling" in *Lawrence Berkeley Report No. LBL-35320*, LBL, California, 1994.

Landers, R.R. *Man's Place in the Dybosphere*. Eaglewood Cliffs, N.J.: Prentice-Hall, 1966.

LaPorte, C.F., et al. *The Earth and Human Affairs*, Committee on Geological Sciences, Division of Earth Sciences, ARC-NAS. San Francisco: Canfield Press, 1972.

Laune, Ian C., ed. *Nature in Cities*. New York: John Wiley, 1979.

Lawrence Berkeley National Laboratory, pamphlet, 1993.

Lawrence Berkeley National Laboratory, pamphlet, 1996.

Lawson, B. *Building Materials, Energy and the Environment: Towards Ecologically Sustainable Development*. Solarch, School of Architecture, University of New South Wales, RAIA, 1996.

Leontief, W. "Environmental Repercussions and the Economic Structure: An Input-Output Approach." *Review of Economics and Statistics*, LII, (August 1970), pp. 262–71.

Leopold, L.B., F.E. Clarke, B.B. Harshaw, and J.R. Balsey. *A Procedure for Evaluating Environmental Impact*. Geological Survey Circular, no. 645, U.S. Geological Survey. Washington, D.C.: U.S. Dept. of the Interior, 1971.

Lloyd Jones, D. *Architecture and the Environment. Bioclimatic Building Design*. London: Laurence King Publishing, 1998.

Lovejoy, D. "The Need for Landscape Planning." RTPI/ILA one-day seminar, Nottingham, Nov. 21, 1973.

Lovins, A.B. *Soft Energy Paths*. New York, Marmondsworth: Penguin Books, 1977.

MAB. *Expert Panel on the Rule of Systems Analysis and Modelling Approaches in the Man and Biosphere Programmes*. Paris, April 18–20, 1972.

McHale, J. "World Dwelling." *Perspecta, The Yale Architecture Journal*, 12, 1967, pp. 120–129.

McHale, J. *The Ecological Context*. New York: Braziller, 1970.

McHarg, Ian. *Design with Nature*. Garden City, NY: Natural History Press, 1969.

McHarg, I., et al. *Amelia Island, Florida. A Report on the Master Planning Process for a New Recreational Community*. Philadelphia: WMRT, 1971.

McHarg, I. *Hazleton: An Ecological Planning Study*. Philadelphia: Dept. of Landscape Architecture and Regional Planning, Graduate School of Fine Arts, University of Pennsylvania, 1973.

McHarg, I., et al. *Pardisan Park in Tehran. A Feasibility Study for an Environmental Park in Tehran, Iran*. Philadelphia: The Mandala Collaborative, 1973.

McKibben, Bill. *The End of Nature*. New York: Random House, 1989.

MacKenzie, D. *Green Design, Design for the Environment*. London: Laurence King Publishing, 1997, p. 8.

McKillop, A. Poster for "Low Impact Technology," 1972.

McLaren, V.W. "Urban Sustainability Reporting." *Journal of the American Planning Association*, vol. 62, no. 2, 1996, p. 84.

Madison, Cathy. "The Hidden Power of Houseplants." *Utne Reader*, May-June 1998, LENS Publishing Co., p. 88.

Makhijani, A.B., and Lichtenberg, A.J. "Energy and Well-Being." *Environment*, vol. 14, no. 5, June 1972.

Marras, Amerigo, ed. *ECO-TEC, Architecture of the In-Between*. New York: Princeton Architectural Press, 1999.

Martin, L. "Architects' Approach to Architecture: Sir Leslie Martin." *RIBAJ*, May 1967, vol. 74, 1967, pp. 191–200.

Mascaro, L., G. Dutra, and F. Finger. "Environmental Aspects of the Urban Precincts in a Subtropical City" in *Environmentally Friendly Cities, Proceedings of PLEA '98*, Lisbon, Portugal, June 1998. London: James & James Science Publishing Ltd., 1998, pp. 99–102.

Maser, Chris. *The Redesigned Forest*. San Pedro, Calif.: R. & E. Miles, 1988.

Matthews, W.H., F.E. Smith, and E.D. Golberg, eds. *Man's Impact on Terrestrial and Oceanic Ecosystems*. Cambridge: MIT Press, 1971.

Meadows, D.H., and D.L. Meadows. *The Limits to Growth*. New York: University Books, 1972.

Mellanby, K. *The Biology of Pollution*. London: Edward Arnold, 1972.

Miller, J.G. "Living Systems: Basic Concepts." *Behavioural Science*, vol. 10, Oct. 1965, pp. 193–237.

Miller, Ronald E. "Interregional Feedback Effects in Input-Output Models: Some Preliminary Results." *Papers & Proceedings of the Regional Services Association*, vol. 17, 1966, pp. 105–125.

Moorcroft, C., ed. "Designing for Survival." *Architectural Design*, vol. XLII, July 1972, pp. 413–445.

Moorcroft, C. "Some Proposals for the Reserving of an Urban Terrace House." *Street Farmers*, 1973.

Moore, Fuller. "Concepts and Practice of Architectural Daylighting" in *Fenestration R & D*. Waiter, Lawrence Berkeley National Laboratory Energy & Environmental Division Publishing, 1996, p. 9.

Morgan, W. "Earth Architecture: Up to Earth." *Progressive Architecture*, April 1979, pp. 84–7.

Morrison, W.I. "The Development of an Urban Interindustry Model: 1. Building Input-Output Accounts." *Environment and Planning*, vol. 5, 1972, pp. 369–385.

National Academy of Sciences: *One Earth, One Future*. Washington, D.C.: National Academy Press, 1990.

Nicholson, M. *The Environmental Revolution*. New York: McGraw-Hill, 1970.

Nihon Sekkei. *Energy Conservation Measures for High-Rise Buildings*. Tokyo: JETRO (Japan External Trade Organisation) Publishing, 1984.

Nikken Sekkei, Pamphlet, 1996.

Norton, G.A., and J.W. Parlour. "The Economic Philosophy of Pollution: A Critique." *Environment and Planning*, vol. 4, 1972, pp. 3–11.

Odum, E.P. "Relationship Between Structure and Function in the Ecosystem." *Journal of Ecology*, vol. 12, 1962, pp. 108–118.

Odum, E.P. *Ecology*. New York: Holt, Rienhart, Winston, 1963.

Odum, E.P. "The Strategy of Ecosystem Development." *Science*, vol. 164, April 1969, pp. 262–270.

Odum, E.P. *Fundamentals of Ecology*. Philadelphia: W.B. Sanders & Co, 1971.

Odum, E.P. "Ecosystems" in W. White and F.J. Little, eds., *North American Reference Essays of Ecology and Pollution*. Philadephia: North American Publishing Co., 1972, pp. 66–69.

Odum, H.T. *Environment, Power and Society*. New York: Wiley-Interscience, 1971.

Odum, H.T., and L.L. Peterson. "Relationship of Energy and Complexity in Planning." *Architectural Design*, October 1972, pp. 624–628.

Ogburn, C. "Where the Food is to Come From." *Population Bulletin*, June 2, 1970, p. 8.

Olgyay, Victor. *Design with Climate*. Princeton, N.J.: Princeton University Press, 1993.

O'Riordan, T. *Perspectives on Resource Management*. London: Pion Ltd., 1971.

Ortega, A., W. Rybezynoki, S. Ayord, W. Ali, and A. Acheson. *The Ecol Operation*. Montreal: Minimum Housing Group, School of Architecture, McGill University, 1972.

Ovington, J.D. "The Ecological Basis of the Management of Woodland Nature Reserves in Great Britain." *Journal of Ecology*, vol. 52 (suppl.), 1964, pp. 29–37.

Paehlke, R. *Environmentalism and the Future of Progressive Politics*. New Haven: Yale University Press, 1989.

Papanek, V. *Design for the Real World: Human Ecology and Social Change*. Chicago: Academy Chicago Publishers, 1985.

Papanek, V. *The Green Imperative, Natural Design for the Real World*. London: Thames & Hudson, 1995.

Patterson, W.C. *The Energy Alternative: Changing the Way the World Works*. London: Optima, MacDonald & Co. Ltd., 1990.

Pawley, M. "The Cambridge Autonomous House." *Building Design*, November 1974.

Pearson, D. *The Natural House Book*. New York: Simon & Schuster Inc., 1989.

Penportier, B., and P.D. Redregal. "Application of Life-Cycle Simulation to Energy and Environment Conscious Design" in *Environmentally Friendly Cities, Proceedings of PLEA '98*, Lisbon, Portugal, June 1998. London: James & James Science Publishing Ltd., 1998, pp. 517–520.

Peranio, A. *The Environmental Crisis–A Cybernetic Challenge*. Haifa, Israel: Technion, 1973, pp.122.

Permasteelisa, pamphlet, 1998.

Pielou, E.C. *An Introduction to Mathematical Ecology*. New York: Wiley-Interscience, 1969.

Plass, G.N. "Carbon Dioxide and Climate." *Scientific American*, July 1969, p. 6.

Poore, M.E.D. "Ecology and Conservation in Land-use." *Chartered Surveyor*, November 1972.

Raju, M.K. "Success Stories in Energy Conservation as Relevant to Thailand" in *Proceedings of the Conference on Energy Conservation: Thailand Means Business*, NRG-CON, Aug. 22-23, 1996, Thailand.

Ramphal, Shridath. *Our Country, the Planet*. Washington, D.C.: Island Press, 1992.

Ray, C. "Ecology, Law and the Marine Revolution." *Biological Conservation*, vol. 3, no. 1, 1970.

Redding, M.J. *Aesthetics in Environmental Planning*. EPA-600/5-73-009. Washington, D.C.: Office of Research and Development, E.P.A., U.S. Government Printing Office, Nov. 1973.

Roaf, S., and M. Hancock, eds. *Energy Efficient Building: A Design Guide*. New York: John Wiley & Sons, 1992.

Roberts, D.G.M. "Public Health Engineering in the External Environment: Water Supply and Reuse, Waste Disposal and Pollution Control." *Philosophical Transactions–Physical Sciences and Engineering, Royal Society of London*, A 272, 1972, pp. 639–50.

Robertson, J., Healthy Buildings International Pty Ltd. *Increasing Human Performance by Improving Indoor Air Quality*. Paper (unpubl.), New South Wales, Australia, 1990.

Rosen, H.J. "Energy Crisis and Materials." *Progressive Architecture*, vol. 54, Dec. 1973, p. 76.

Rosenfield, A.H., J.J. Rowan, H. Akbari, and A.C. Lloyd. "Painting the Town White and Green." *Technology Review*, Feb./March 1997, pp. 52–59.

Sagasti, F. "A Conceptual and Taxonomic Framework for the Analysis of Adaptive Behaviour" in *General Systems: Yearbook for the Society*

for the Advancement of General Systems, vol.
XV, 1970, pp. 151–60.

Salem, Osama Shible. "Toward Sustainable Architecture and Urban Design: Categories, Methodologies and Models." Troy, NY: Rensselaer Polytechnic Institute, unpublished manuscript, 1990.

Schudltz, A.M. "A Study of an Ecosystem: The Artic Tundra" in G.M. Van Dyne, ed., The Ecosystem Concept in Natural Resources Management. New York: Academic Press, 1969, pp. 77–93.

Seborg, G.T. "The Environment and What to Do About It." Nuclear News, July 1969.

Seddan, J.W. A Proposal to the National Endowment for the Arts for Support of a Project in Environmental Design. Mimeo, 1973.

Sharkawy, M.A., and J.A. Graaskamp. Inland Lakes Renewal and Management Demonstration. Madison, Wisc.: Environmental Awareness Center School of Natural Resources, University of Wisconsin, 1971.

Shiva, V. Biodiversity, A Third World Perspective. Pulau Penang, Malaysia: Third World Network, 1993.

Simon, H.A. The Sciences of the Artificial. Cambridge, Mass.: MIT Press, 1969.

Simonds, J.O. Earthscape, A Manual of Environmental Planning. New York: McGraw-Hill, 1978.

Sitarz, D., ed. Agenda 21: The Earth Summit Strategy to Save Our Planet. Boulder, Colo.: The Earth Press, 1994.

Sjors, H. "Remarks on Ecosystems." Svensk Botanisk Tidskrift, vol. 49, H 1-2, 1955.

Skinner, B. Earth Resources. Englewood Cliffs, N.J.: Prentice Hall, Inc., 1969.

Slessor, C. Eco-Tech: Sustainable Architecture and High Technology. London: Thames & Hudson, 1997.

S.M.I.C. Inadvertent Climate Modification: Report of the Study of Man's Impact on Climate (SMIC). Cambridge, Mass.: MIT Press, 1970.

Smith, D.R. "Pollution and Range Ecosystems" in W.H. Matthews, F.E. Smith, E.D. Golberg, eds. Man's Impact on Terrestrial and Oceanic Ecosystems. Cambridge, Mass.: MIT Press, 1971.

Socolow, R.H., and J. Harte. Patient Earth. New York: Holt Rinehart and Winston Inc., 1971.

Sorenson, J.C. "Some Procedures and Programs for Environmental Impact Assessment" in R.B. Ditton and T.L. Goodale, eds., Environmental Impact Analysis: Philosophy and Methods. Environmental Impact Analysis, Proceedings of the Conference in Green Bay, Wisc.: Sea Grant Publishing, WIS-SG-72-111, 1972, pp. 97–106.

Sorenson, J.C., and J.E. Pepper. Procedures for Regional Clearinghouse Review of Environmental Impact Statements–Phase Two, draft. Berkeley, Calif.: Association of Bay Area Governments, 1973.

Sorenson, J.C., and M.L. Mass. Procedures and Programs to Assist in the Environmental Impact Statement Process. Los Angeles: University of California, University of Southern California, SG-PUB-No. 27, USC-SG-AS2-73, 1973.

Spellerberg, I.F. Conservation Biology. Harlow, England: Longman Group Ltd., 1996.

Spofford, W.O. "Residuals Management" in W.H. Matthews, F.E. Smith, and E.D. Golberg, eds., Man's Impact on Terrestrial and Oceanic Ecosystems. Cambridge, Mass.: MIT Press, 1971, pp. 477–88.

Spofford, W.O. "Total Environmental Management Models" in R.A. Deininger, ed., Models for Environmental Pollution Control. Ann Arbor, Mich.: Ann Arbor Scientific Publishing Inc., 1973.

St. John, A., ed. The Sourcebook for Sustainable Design. Boston, Mass.: Architects for Social Responsibility, 1992.

Stamp, D. Nature Conservation in Britain. London: Collins, 1969.

Steele, J. Sustainable Architecture: Principles, Paradigms and Case Studies. New York: McGraw-Hill, 1997.

Szczelkun, S.A. Survival Scrapbook 3: Energy. New York: Schocken Books, 1973.

Takeuchi, Kazuhiko, ed. Ecological Landscape Planning. PROCESS: Architecture 127. Tokyo: Process Architecture Co. Ltd., 1995.

Taylor, Peter. Respect for Nature. Princeton: Princeton University Press, 1986.

Tearle, K., ed. Industrial Pollution Control: The Practical Implications. London: Business Books, 1973.

Tebbutt, T.H.Y. Principles of Water Quality Control. Oxford: Pergammon Press, 1971.

Thring, J.B. The Autonomous House Project. Cambridge: Department of Architecture, University of Cambridge, 1973.

Todd, N.J., and J. Todd. Bioshelters, Ocean Arks, City Farming: Ecology as the Basis of Design. San Francisco: Sierra Club Books, 1984.

Todd, N.J., and J. Todd. From Eco-Cities to Living Machines: Principles of Ecological Design. Berkeley, Calif.: North Atlantic Books, 1994.

Toftner R.O. "A Balance Sheet for the Environment." Planning, July 1973.

Tolman, E.C., and E. Brunswick. "The Organism and the Causal Texture of the Environment." Psychological Review, vol. XLII, pp.43–77.

Tubbs, C.R., and J.W. Blackwood. "Ecological Evaluation of Land for Planning Purposes." *Biological Conservation*, vol. 3, no. 3, April 1971, pp. 169–72.

Turk, A., J. Turk, and J. Wittes. *Ecology Pollution Environment*. Philadelphia: W.B. Saunders, Co., 1972.

Turner, Tom. *Landscape Planning and Environmental Impact Design*. London: UCL Press, 1998, p. 95.

U.S.D.H. (U.S. Department of Health). *1968 National Survey of Community Solid Waste Practices*. Cincinnati: U.S. Department of Health, Education and Welfare, Environmental Control Administration, 1968.

Vale, B. *The Automonous House*. Cambridge: Technical Research Division Publ., Department of Architecture, University of Cambridge, 1972.

Vale, B., and R. Vale. *Towards a Green Architecture: Six Practical Case Studies*. London: RIBA Publications Ltd., 1991a.

Vale, B., and R. Vale. *Green Architecture: Design for a Sustainable Future*. London: Thames & Hudson, 1991b.

Van der Ryn, S., and P. Calthorpe. *Sustainable Communities: A New Design Synthesis for Cities, Suburbs and Towns*. San Francisco: Sierra Club Books, 1991.

Van Dyne, G.M. *Ecosystems, Systems Ecology and Systems Ecologists*, ORNL-3957. Oak Ridge, Tenn.: Oak Ridge National Laboratory, 1966.

Van Dyne, G.M., ed. *The Ecosystem Concept in Natural Resource Management*. New York: Academic Press, 1969.

Victor, P.A. *Pollution: Economy and Environment*. London: Allen and Unwin, Ltd., 1972.

Villecco, M., ed. *Energy Conservation in Building Design*. Washington, D.C.: AIA, 1978.

Von Weizsacker, E., A.B. Lovins, and L.H. Lovins. *Factor Four: Doubling Wealth, Halving Resource Use*. London: Earthscan Publications Ltd., 1997.

Waggoner, P.E. "Weather Modification and the Living Environment" in F.F. Darling and J.P. Milton, eds., *Future Environments of North America*. Garden City, NY: The Natural History Press, 1966.

Walmsley, D.J. *Systems Theory: A Framework for Human Geographical Enquiry*, Publ. HG/7. Canberra: Research School of Pacific Studies, Australian National University, ANU Press, 1972.

Walton, K. *The Problem of Developmental Impacts on the Countryside*. Aberdeen: Department of Geography, University of Aberdeen, Mimeo, 1973.

Wann, D. *Deep Design*. Washington, D.C.: Island Press: 1996.

Warren, R. *The Urban Oasis: Guideways and Greenways in the Human Environment*. New York: McGraw-Hill, 1998.

WCED. *Our Common Future*. Oxford: The World Commission on Environment and Development Oxford University Press, 1987.

Weddle, A.E., and J. Pickard. "Techniques in Landscape Evaluation." *Journal of the Town Planning Institute*, vol. 55, no. 9, November 1969.

Weimer, A.D., and H. Hoyt. *Real Estate*. New York: Ronald Press Co., 1966, pp.74

Wells, M.B. *The Great Ecologic Colouring Book of Life and Death and Architecture*. Cherry Hill, N.J.: The Conservation Account, 1971.

Wells, M.B. "An Ecologically Sound Architecture is Possible." *Architectural Design*, July 1972, p. 433.

Wells, M. *Gentle Architecture*. New York: McGraw-Hill, 1984.

Wettqvst, O.F., et al. *Identification and Evaluation of Coastal Resource Patterns in Florida*. Tampa: Florida Coastal Coordination Council, 1971.

White, G.F. "Environmental Impact Statements." *The Professional Geographer*, vol. XXIV, no. 4, November 1972, pp. 302–9.

Wilen, J.E. "A Model of Economic System–Ecosystem Interaction." *Environment and Planning*, vol. 5, 1973, pp. 409–20.

Willard, B.E., and J.W. Marr. "Effects of Human Activities on the Alpine Tundra Ecosystem in Rocky Mountain National Park, Colorado." *Biological Conservation*, vol. 2, no. 2, July 1970.

Williams, E.R., and P.W. House. *The State of the System Model (SOS): Measuring Growth Limitations Using Ecological Concepts*, U.S. Environmental Protection Agency, Socio-Economic Environmental Studies Series, EPA-600/5-73-013. Washington, D.C.: Office of Research and Development EPA, U.S. Government Printing Office, February 1974.

Windheim, L.S. "A System Morphology for Examining Energy Utilisation in Buildings." *J.BRAB Building Research Institute*, NAS, July/December 1973, pp.1–6.

Wolverton, B.C. *Eco-Friendly Houseplants*. London: Weidenfield & Nicolson, 1996.

Woodwell, G.M. "Effects of Pollution on the Structure and Physiology of Ecosystems" in S.W.H. Matthew, F.E. Smith, and E.D. Golberg, eds., *Man's Impact on Terrestrial and Oceanic Ecosystems*. Cambridge, Mass.: MIT Press, 1971, pp. 47–58.

Woodwell, G.M., and C.A.S. Hall. "The Ecological Effects of Energy: A Basis for Policy in Regional Planning" in *Energy, Environment and Planning in the Long Island Sound Region.* Upton, NY: Brookhaven National Labs., 1972.

Woolley, T., et al. *Green Building Handbook: A Guide to Building Products and their Impact on the Environment.* London: E. & F.N. Spon, 1997.

World Resources Institute. *World Resources 1994–95.* New York: Oxford University Press, 1994.

Wott, K.E.F. *Ecology and Resource Management: A Quantitative Approach.* New York: McGraw-Hill, 1968.

Yeang, K. "Bases for Ecosystem Design." *Architectural Design,* July 1972, pp. 434–6.

Yeang, K. *Whole Earth Energy and Materials Catalogue.* London: AA Publ. for Unit 5, 1974a.

Yeang, K. "Energetics of the Built Environment." *Architectural Design,* July 1974b, pp. 446–51.

Yeang, K. "Bionics–The Use of Biological Analogies for Design." *Architectural Association Quarterly (AAQ),* vol. 6, no. 2, 1974c.

Yeang, K. *Bioclimatic Skyscrapers.* London: Ellipsis London Ltd., 1994.

Yeang, K. *Designing with Nature: The Ecological Basis for Architectural Design.* New York: McGraw-Hill, 1995.

Yeang, K. *The Skyscraper Bioclimatically Considered: A Design Primer.* London: Academy Editions, 1996.

Yeang, K. *T.R. Hamzah & Yeang Selected Works.* Australia: The Images Publishing Group Pty Ltd., 1998, p. 7.

Zeiher, L.C. *The Ecology of Architecture: A Complete Guide to Creating the Environmentally Conscious Building.* New York: Whitney Library of Design, 1996.

Index

Front cover: Tokyo-Nara Building
Back cover: Editt Tower (photograph), BATC (drawing), and UMNO Building
(wind rose sketch)

© Prestel Verlag, Munich · London · New York, 1999
All illustrations courtesy of the author, by prior arrangement between the
author and the respective copyright holders where applicable. These sources
are listed in the captions next to the illustrations as they appear in the book.

Prestel Verlag
Mandlstrasse 26, D-80802 Munich, Germany
Tel. +49 (89) 38 17 09-0, Fax +49 (89) 38 17 09-35
4 Bloomsbury Place, London WC1A 2QA
Tel. +44 (0171) 323-5004, Fax +44 (0171) 636-8004
and 16 West 22nd Street, New York, NY 10010, USA
Tel. (212) 627-8199, Fax (212) 627-9866

Prestel books are available worldwide.
Please contact your nearest bookseller or write to any
of the above addresses for information concerning
your local distributor.

Library of Congress Cataloging-in-Publication Data

Yeang, Ken. 1948-
  The green skyscraper : the basis for designing sustainable intensive buildings /
  Ken Yeang
  p. cm.
Includes bibliographical references.
ISBN 3-7913-1993-0 (alk. paper)
1. Skyscrapers--Environmental aspects I. Title.
NA6230.Y43 1999
720'. 47--dc21
98-34625
CIP

Edited by Bruce Murphy
Designed by Verlagsservice G. Pfeifer, Germering
Typesetting by EDV-Fotosatz Huber, Germering
Lithography by Repro Line, Munich
Printed and bound by Bosch Druck, Landshut

Printed in Germany on acid-free paper
ISBN 3-7913-1993-0